RELATIVITY AND QUANTUM PHYSICS

FOR BEGINNERS®

RELATIVITY AND QUANTUM PHYSICS

FOR BEGINNERS®

Steven L. Manly
Illustrated by Steven Fournier

FOR BEGINNERS®

Published by For Beginners LLC
155 Main Street, Suite 211
Danbury, CT 06810 USA
www.forbeginnersbooks.com

A For Beginners® Documentary Comic Book

Cataloging-in-Publication information is available from the Library of Congress.

ISBN # 978-1-934389-42-3 Trade
Manufactured in the United States of America

For Beginners® and Beginners Documentary Comic Books® are published by For Beginners LLC.

Reprint Edition
10 9 8 7 6 5 4 3 2

This book is dedicated to you, the reader. It is my hope that it brings you enjoyment and a deeper appreciation of nature's strange reality show.

CONTENTS

Chapter 1

Science and the Human Bias

Humans seek to understand the universe!

Do you BELIEVE?
By golly, I see yet ANOTHER star!
Ah, now. The distant look in the eyes shows the determined yearning of man to be liberated from the bonds of oppression. I can enhance that with just a touch of a blue in the lips I think . . .
The meaning of life is pudding.
Pudding?! Whoa . . . you're a freak, dude. Waves and chicks, man. That's what floats my board.

Religion is based on faith. Art is based on aesthetics. While both religion and art can provide insight into the human condition, the methodology of science is unique in that it bows to observations. Ideas that are not consistent with what we see in nature under controlled and repeatable circumstances are *thrown out*!

> "Faith" is a fine invention
> When Gentlemen can see,
> But Microscopes are prudent
> In an Emergency.
> — Emily Dickinson

"Methodology" of science? Say WHAT?
You see, Gertrude? I keep telling you the geek shall inherit the Earth!
After you inherit the Earth, maybe you two can afford to get a room?

In science a person looks at something and makes a hypothesis (or theory) about how it works. Then they design an experiment to test the hypothesis. After doing the experiment, the person modifies or discards the theory depending on the results of the experiment. This process repeats, and our scientific understanding of the phenomenon evolves.

My second-grade teacher, Mrs. Broccoli, called this the *scientific method.*

Communication, honesty, and reproducibility of observations lie at the core of what makes science work. Experimental results must be conveyed to others unambiguously and in detail so that others can reproduce the experiment.

I tell you, in my microscope the red doohicky is much smaller than the blue thingumajig.
BLUE ONE
That's strange. My observations show the red one is definitely larger than the blue one.
RED ONE

And this makes typical scientific writing **BORING** and **DRY.**

If you don't believe that, read this:

> **This paper describes the measurement of the energy dependence of elliptic flow for charged particles in Au+Au collisions using the PHOBOS detector at the Relativistic Heavy Ion Collider (RHIC). Data taken at collision energies of $\sqrt{s_{NN}}$ = 19.6, 62.4, 130 and 200 GeV are shown over a wide range in pseudorapidity. These results, when plotted as a function of $\eta' = |\eta| - y_{beam}$, scale with approximate linearity throughout η', implying no sharp changes in the dynamics of particle production as a function of pseudorapidity or increasing beam energy.**

But I think this writing is very exciting!

This is why it seems like you have to go to school forever to understand this crap.

There's no room for ambiguity and confusion in scientific communications. This leads to a precise, layered, specialized language—or lingo—in each area of science.

Also, this desire for unambiguous clarity, along with the basic quantitative nature of many measurements, leads to the heavy use of mathematics in science.

> Mathematics is a more powerful instrument of knowledge than any other that has been bequeathed to us by human agency. —René Descartes

Music communicates . . . but it evokes different feelings in different people.

Mathematics and very precise language allow scientists to communicate with as little confusion as possible.

Yeah? Well it sure as heck confuses the rest of us!

There's more to it than clarity. Mathematics and layers of concepts often lead to the ability to ask questions and have insights that are not possible otherwise.

So, Herr Professor, tell me ... a scientist equipped with a well-constructed theory backed up by the right lingo and mathematics MUST ALWAYS be right! Isn't that so?

Well, no. the realm of absolute truth is religion and not science. A scientific idea is always vulnerable to being overthrown or modified as new things are discovered about the universe. Also, scientists are human. This means there is a human bias to everything they do.

> The most remarkable discovery made by scientists is science itself. —Gerard Piel

Science geeks have feelings too, you know.

What is the human bias?

Natural human tendencies

Simple mistakes

In spite of scientists' attempts to make unambiguous measurements, human judgment and intuition often comes into play.

Scientists sometimes stop looking for errors in an experiment or data analysis when they get the answer they expect to find, yet they look very hard for problems if they see something unexpected.

> An expert is a man who has made all the mistakes which can be made in a very narrow field. — Niels Bohr

The limitation of experience

Our senses, intuition, and tendency to interpret data are tuned to times and distances and speeds and sizes that are commonly encountered. This is what we know. Our expectations are biased toward the realm of our experience. Every time we create a new technology that allows us to see farther, smaller, or faster things we are forced to expand our minds to encompass the unexpected.

Anthropocentric and geocentric ideas

Humans have always wanted to feel important and have tended to like ideas that place them at the center of the universe. Religions often play to this desire.

It is SO all about ME!
AND me, right?

Nature doesn't seem to have the same hang-up.

Overcoming the human bias

The methodology of science tends to push us beyond the human bias.

Experimental results are shared and experiments are repeated. This leads to the constructive and frank interchange of ideas and the correction of earlier mistakes.

I am NOT wrong!
Oh yes you are! You're so dumb that your ancestors were used as the control group during evolution!

A hypothesis with a human bias is fair to propose. After all, perhaps we *are* special!

But, in science, that hypothesis (just like any other) must be supported by experimental data if it is to survive. Scientists tend to prefer simpler explanations when given a choice and all other things are equal.

As strange as it may seem, aesthetics does have a place in science. Science has an artistic side. A critical part of the methodology of science is known as Ockham's razor—when choosing between different theories that describe the data, the simplest is often the best.

> Numquam ponenda est pluralitas sine necessitate.
> (Plurality is never to be posited without necessity.)
> —William of Ockham

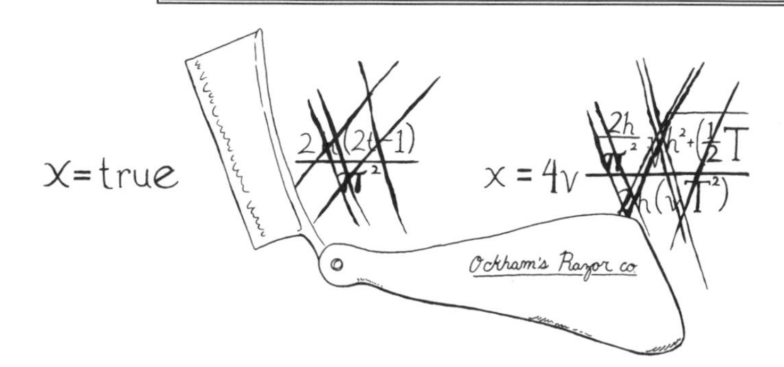

x=true
Ockham's Razor co

The progression of Man's view of earth's place in the cosmos is a good example of science overcoming the human bias.

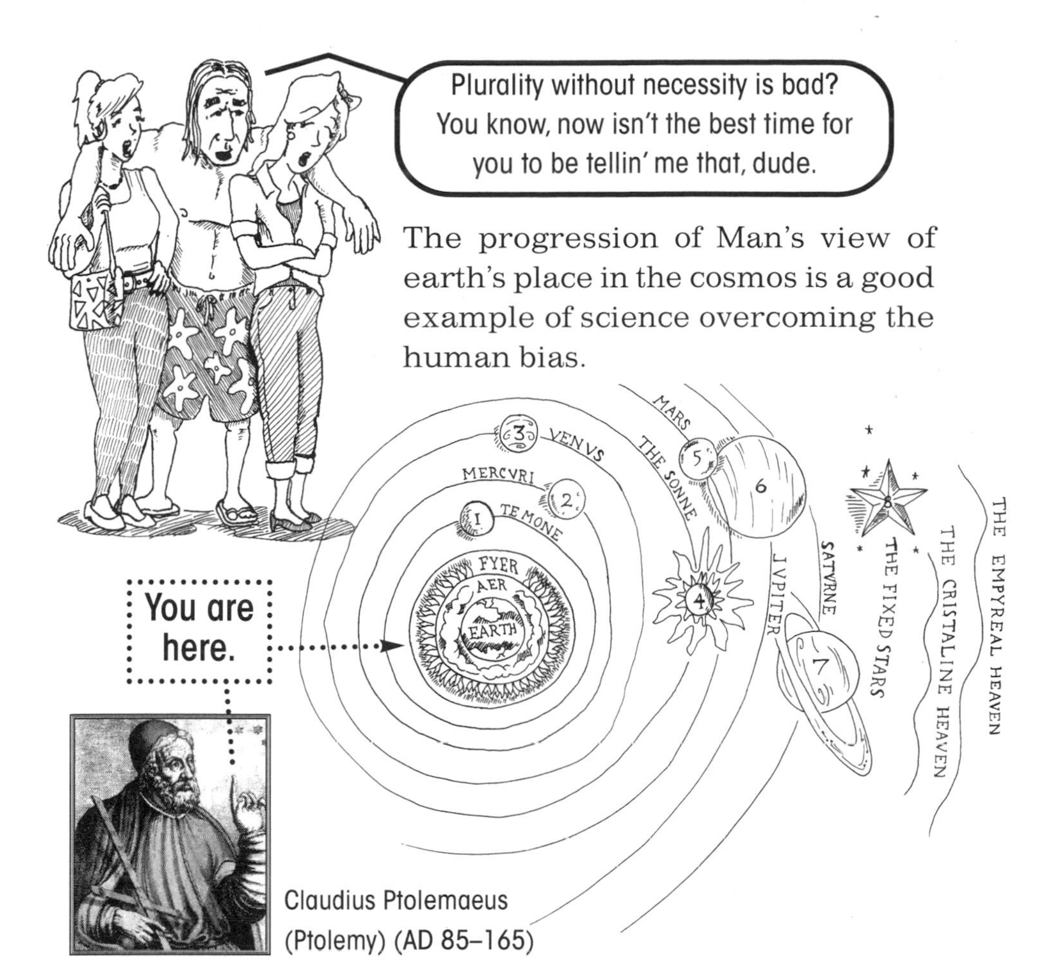

Plurality without necessity is bad? You know, now isn't the best time for you to be tellin' me that, dude.
You are here.
MARS
VENVS
THE SONNE
MERCVRI
TE MONE
FYER
AER
EARTH
1
2
3
4
5
6
7
JVPITER
SATVRNE
THE FIXED STARS
THE CRISTALINE HEAVEN
THE EMPYREAL HEAVEN

Claudius Ptolemaeus (Ptolemy) (AD 85–165)

The most widely held view of cosmology through the Middle Ages was that catalogued by the Egyptian astronomer Ptolemy in his book *Almagest,* written in AD 150. The Ptolemaic universe contained many elements proposed long before his time by others, such as Aristotle and followers of Pythagoras. In Ptolemy's view, the sun, moon, stars, and five known planets move around the earth on a complex system of nested, rotating transparent (or crystalline) spheres and circles within circles. Because the planets, sun, and moon each have unique motions in the sky relative to the stars, the complicated multishell and circle-within-circle arrangement was necessary in order for the model to agree with observations of the heavenly bodies.

Then came along a Prussian astronomer (born in what is now Poland), named Nicolaus Copernicus (1473–1543), and he . . .

I know this one! Let me tell it. ... Copernicus looked at that Ptolemy dude's crazy drawing and said, "Man, this is way too complicated. Check this out. If you put the sun in the center instead of the earth, things are simpler. You know . . . no plurality without necessity and all that crap."

As he neared death, Copernicus published *De revolutionibus orbium coelestium (On the Revolutions of the Celestial Spheres)*, which presented a heliocentric (sun-centered) view of the universe that was eventually shown to be simpler than Ptolemy's cosmology.

Humans less important (moved from center of universe)

Simplicicity (Ockham's razor in action)

A man born shortly after the death of Copernicus, Danish astronomer Tycho Brahe, made very careful measurements of the motion of the heavenly bodies—much more accurate and precise than were available before. Brahe created a cosmological model where the sun and the moon moved in circles about the earth while the planets moved in circles about the sun.

Technology
improves
observations

Old models inconsistent with observations thrown out as new model brought in

Armed with Brahe's data, a German astronomer and mathematician named Johannes Kepler—who had been an assistant to Brahe and "appropriated" his data upon Brahe's death—pursued his study of Brahe's data and his own observations and

To create new model, the fundamental biases long held dear are overthrown (e.g., movement of planets in circles and spheres tossed out because ellipses fit the data)

discovered that a heliocentric system with the planets having slightly elliptical orbits was best able to describe the data. He developed three laws of planetary motion that eventually were explained by Isaac Newton and his theory of gravity.

Johannes Kepler (1571–1630)

THE EARTH IS THE CENTER OF THE UNIVERSE.
THE UNIVERSE IS HELIO-CENTRIC.
UH, GUYS? YOU'LL WANT TO LOOK AT THIS.
PTOLEMAEOUS
GALILEO
KEPLER

The final blow against the Ptolemaic/geocentric universe came in 1610, when an Italian named Galileo Galilei used a new device—a telescope—to observe the phases of Venus. The exhibition of phases by Venus was strong evidence that Venus orbited the sun.

> To know that we know what we know, and to know that we do not know what we do not know, that is true knowledge. —Copernicus

The search for simplicity and consistency in experimental observations coupled with more and better observations allowed mankind to overthrow deeply held convictions about the structure of the universe.

The struggle against the human bias continues to this day. As we have expanded our horizons to see things vastly smaller/faster/larger/farther than ever before, we have been forced to confront preconceptions born of the human experience and create wholly new ways of looking at the world around us. There is nothing quite as strange and exciting as the reality show of our universe.

This book describes the crazy, revolutionary theories of relativity and quantum physics and shows how these ideas have led to amazing advances in our understanding of the universe.

It's a touching little story dah-ling. But, it will never sell. Throw in some gratuitous sex and mindless violence and get back to me.

Tycho Brahe lost much of his nose dueling with swords, and Kepler's mother was tried as a witch.

Chapter 2
Surviving a Trip to the Mall

Oh yeah. I hear there's a sale at the mall! It's *time* to save some money.
Er . . . yeah. Whatever. All I really need are my credit cards.

Headed to the mall for the perfect accessory or that new pair of running shoes? To get there and find that perfect thing, you'll need a fundamental concept of space and time.

> Most institutions demand unqualified faith; but the institution of science makes skepticism a virtue. — Robert K. Merton, *Social Theory and Social Structure* (1962)

The methodology of science relies on people making observations of nature and relating those observations to others. To do that scientists need to have a concept of space and time.

My girlfriend says I'm spacey. Does that count?

The concept of space and time is *not* just for scientists. Shopping, playing soccer, hunting deer, working, even . . . er . . . surfing require that people have a concept of space and time.

Dude. Surfing IS work.

You mean like *space* to put our new stuff from the mall and *time* to enjoy it?
What do you mean exactly when you say "space"?
I know! Space is the final frontier, right? I heard that once on TV.
What is time? Like, I have a watch. But that isn't time.
Space is the fabric in which we measure *where* things and events are located.
Time is the fabric in which we measure *when* things and events are located.
Since "things" around us occupy space, the need for a concept of space is relatively easy to accept. But, how do we know time exists? Why do we need time?
According to modern astronomers, space is finite. This is a very comforting thought—particularly for people who cannot remember where they left things.
—Woody Allen

Time is but the stream I go a-fishing in.
—Henry David Thoreau

Time goes, you say? Ah no!
Alas, time stays, we go.
—Henry Austin Dobson

Time is but the stream I go a-fishing in.
—Henry David Thoreau

Time goes, you say? Ah no!
Alas, time stays, we go.
—Henry Austin Dobson

As you can see, life would be awfully boring without change. Fortunately, the universe around us is not static. In fact, the only constant is change, and change requires the concept of time. Time is the ruler against which change is measured.

The only thing that stays the same is change. —Melissa Etheridge, "Change"
Sweet! Hang ten, Father Time!
The only reason for time is so that everything doesn't happen at once. —Albert Einstein
I view the universe around me in terms of three infinite flat spatial dimensions moving steadily through time in one direction.
Er . . . yeah, that's what I was thinking alright.

All of us share a basic concept of space and time, which is integral to how we view the world.

Why *three* dimensions?

It takes three numbers to specify the position of something in everyday life. Suppose you go to the grocery store looking for corn nuts and you have trouble finding them. Finally you ask the manager for help. In order to lead you to the corn nuts, the manager has to specify the aisle, how far down the aisle you should go, and the shelf on which you should look. That's three numbers, one corresponding to each spatial dimension in which we live. The room that you are sitting in has a length, width, and height—three numbers.

> How long a minute is, depends on which side of the bathroom door you're on.
> —Zall's Second Law

Why *only* three dimensions?
Why not more?
ΦBK

We might live in a universe with more than three spatial dimensions in spite of the fact that we can only perceive three dimensions. How can this be? Imagine being an ant on a large beach ball or a sailor on the ocean. In both cases, the relevant world seems flat and two dimensional. Yet we know both the sailor and ant are moving on a large three-dimensional object. It might be the case that the universe has more to it than meets the eye.

> Space is to place as eternity is to time.
> — Joseph Joubert

Three spatial dimensions moving lockstep through time. This is our shared view of the world. Let's be very clear about what this means. If we handed out synchronized watches to ten different people in a room and asked them to leave, go about their business, but return in exactly one hour, each person would return to the room at the same time, regardless of what they did during that hour. Time is absolute. It moves along at the same rate no matter who you are or what you are doing.

I take exception to that statement, sir. When I go to a robotics convention, time just seems to fly.

Are you kidding me? Like, who goes to robotics conventions?

> Today is the tomorrow we worried about yesterday.
> —Unknown source

To be sure, we perceive time to pass at different rates depending on whether we are having fun or are bored or in pain or in ecstasy. But if we look at a clock, time passes at the same rate for everyone—no matter whether he or she is happy, sad, or indifferent.

What? You mean it's only been an hour?! Sigh.
This sure is fun! I can't believe an hour has passed already!

Oh yeah. This here is my core inner philosophy about time. I heard it on a TV show once. You ready? Here it is: "You can't change the past but you can ruin the present by worrying about the future." Sweet! Is it not?

If you want to learn about space and time, a great place to begin is in the study of motion. Speed is a measure of how far something goes (in space) in a given amount of time. Since speed is a quantity that involves both space and time, our well-defined human intuition about space and time leads us to have particular expectations about speeds.

Speed and velocity are not *quite* the same thing. In everyday language, most of us use the words "speed" and "velocity" interchangeably. Formally, they are slightly different beasts. Velocity is speed with the direction specified. A car can have a velocity of 10 miles per hour north or 10 miles per hour east. In both cases the speed is 10 miles per hour, but the velocities differ because the directions are different.

N W S E

To learn about speed, imagine taking a trip to the mall and sitting at a café watching people on a moving sidewalk. They look at you rather strangely when you pull out the radar gun and start to make speed measurements.

RADAR
STAY BACK, JIMMY. THAT GUY LOOKS A LITTLE STRANGE.
EASTSIDE MALL

What can you learn by watching people and recording speeds?

⇒4mph
⇒2 mph
⇒6mph

You see that a person moves past you on the moving sidewalk at a speed equal to that of his walking speed plus the speed of the sidewalk.

The result from the mall sidewalk radar gun experiment should not be too surprising. In everyday life, velocities add. You see this all around you every day. Want another example? Imagine you are going to the mall in a car that is moving with a speed of 30 miles per hour. Suppose you approach another car from behind that is moving only 25 miles per hour. The fact that velocities add means that your car approaches the other car with a relative speed of 5 miles per hour.

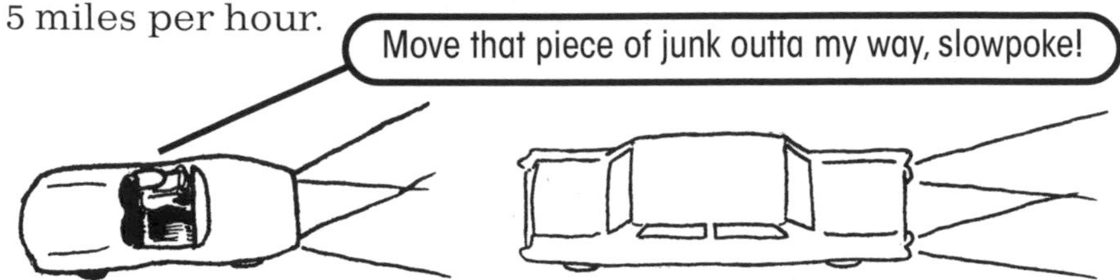
Move that piece of junk outta my way, slowpoke!

If you are driving a car or shooting at a running deer or playing a sport your brain is processing relative velocities constantly. Hunters and football quarterbacks "lead" their targets. The concept of relative velocities that everyone uses is exactly what you measured at the mall in the sidewalk thought experiment. It makes sense to us. It works.

> Nobody in football should be called a genius. A genius is a guy like Norman Einstein.
> —Joe Theismann, former NFL quarterback

12
JOE NAMATH
~ THE EARLY YEARS ~

> I want to rush for 1,000 or 1,500 yards, whichever comes first.
> —George Rogers, former NFL running back

Chapter 3
Nature's Relatively Strange Reality Show

And stay out!

After the mall security chaps decide that your radar gun experiments by the moving sidewalk are freaking out the shoppers, you take your inquisitive nature and radar gun out to the mall parking lot. Measuring the speed of passing cars quickly becomes boring. So you decide to see if you can measure the speed of the *light* emitted by the cars' headlights.

It turns out that a radar gun *can't* measure the speed of light. But don't let that bother you for the moment. Suppose that you *can* measure the speed of the light emitted by a car headlights. What do you see?

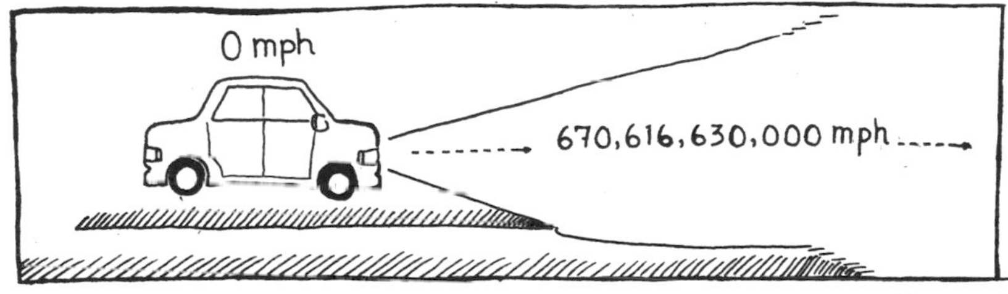
0 mph
670,616,630,000 mph

If the car is not moving, you observe the car emitting light that moves at a speed of 670,616,630,000 miles per hour. You measure the same speed no matter what direction the car is pointed.

For comparison, consider this: F-16 fighter jets can travel at a top speed of approximately 1,500 miles per hour.

The speed of light is even faster than Surfer Dude on a date.

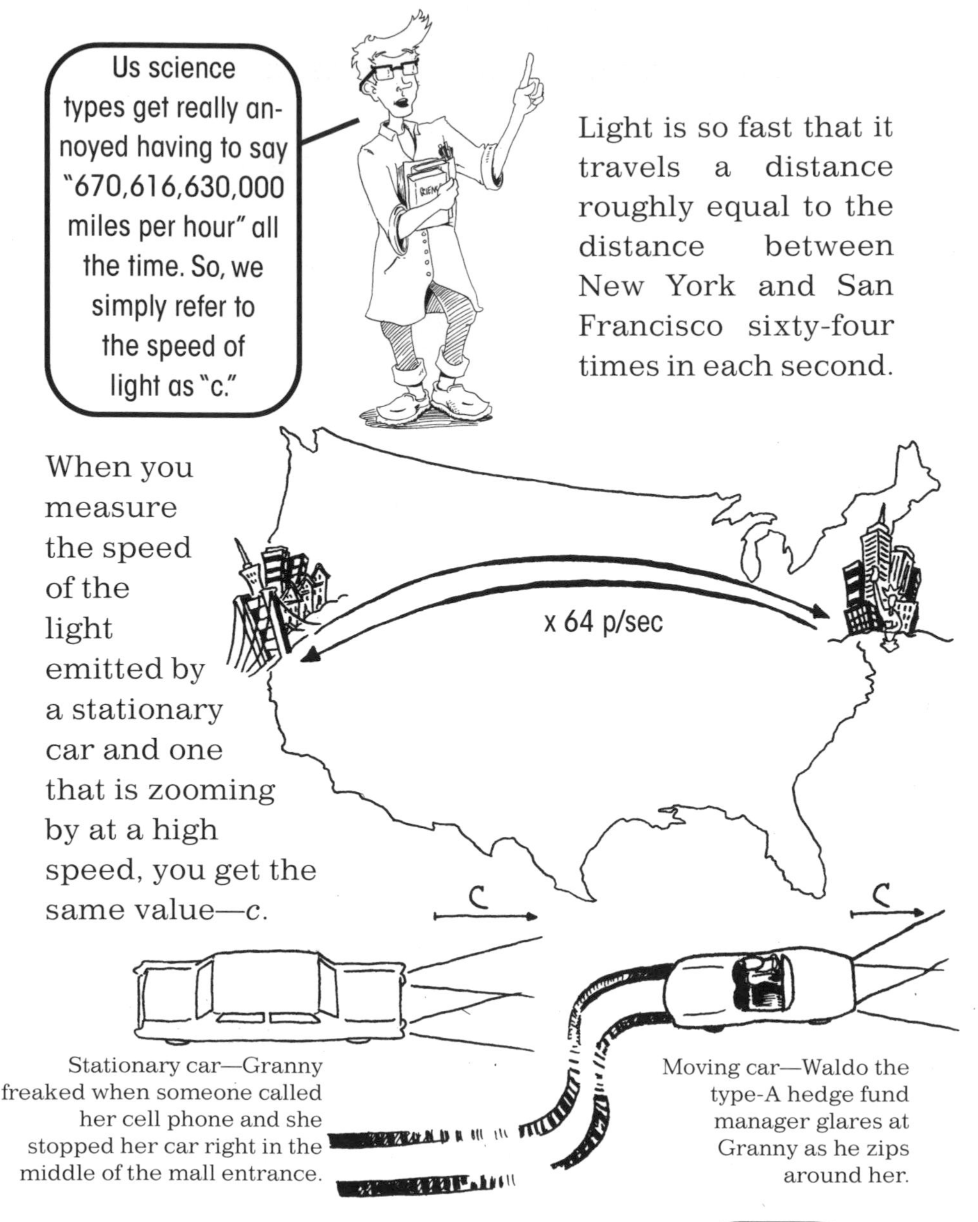
Us science types get really annoyed having to say "670,616,630,000 miles per hour" all the time. So, we simply refer to the speed of light as "c."
x 64 p/sec
c
c

Light is so fast that it travels a distance roughly equal to the distance between New York and San Francisco sixty-four times in each second.

When you measure the speed of the light emitted by a stationary car and one that is zooming by at a high speed, you get the same value—*c*.

Stationary car—Granny freaked when someone called her cell phone and she stopped her car right in the middle of the mall entrance.

Moving car—Waldo the type-A hedge fund manager glares at Granny as he zips around her.

This is the same thing as saying two people walking at the same speed in the mall continue to move at the same speed even after one of them steps onto the moving sidewalk without changing their stride! This violates the common human intuition about relative velocities. *It makes no sense.*

This is completely wacko!

Believe it or not, the fact that light moves at the same speed no matter how you or the source of the light move with respect to each other was not discovered in a mall parking lot by bored science geeks.

Really?! Phooey. I thought I was going to be famous.

RADAR

Albert Michelson and Edward Morley discovered the surprising constancy of the speed of light in 1887 using a device called an interferometer. In this device, a light beam is split and sent in two different directions and recombined. If the light moved at different speeds in the two directions, the brightness of the light would change when the two beams of light recombine due to interference. Michelson and Morley mounted their interferometer on something moving very fast and compared the speed of the light moving along the direction of motion with the speed of light moving at right angles to the direction of motion . . . and they found no difference.

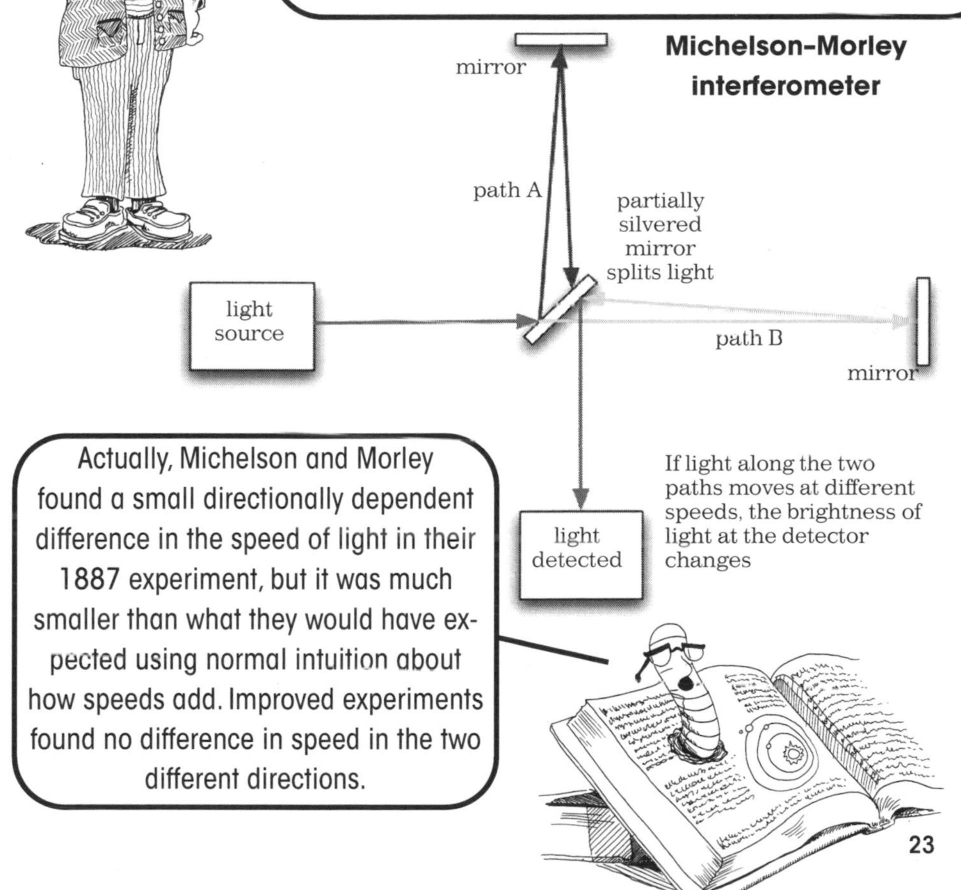

Michelson–Morley interferometer
mirror
path A
partially silvered mirror splits light
light source
path B
mirror
light detected
If light along the two paths moves at different speeds, the brightness of light at the detector changes
Actually, Michelson and Morley found a small directionally dependent difference in the speed of light in their 1887 experiment, but it was much smaller than what they would have expected using normal intuition about how speeds add. Improved experiments found no difference in speed in the two different directions.

The "something fast" that Michelson and Morley used for their experiment—their moving sidewalk—was the earth itself. Earth moves around the sun at a speed of approximately 30 kilometers per second and the solar system moves around the center of the Milky Way galaxy at a speed of roughly 250 kilometers per second and so forth. They compared the speed of light along the direction of Earth's motion to the speed perpendicular to the direction of earth's motion through space.

Wheee!!

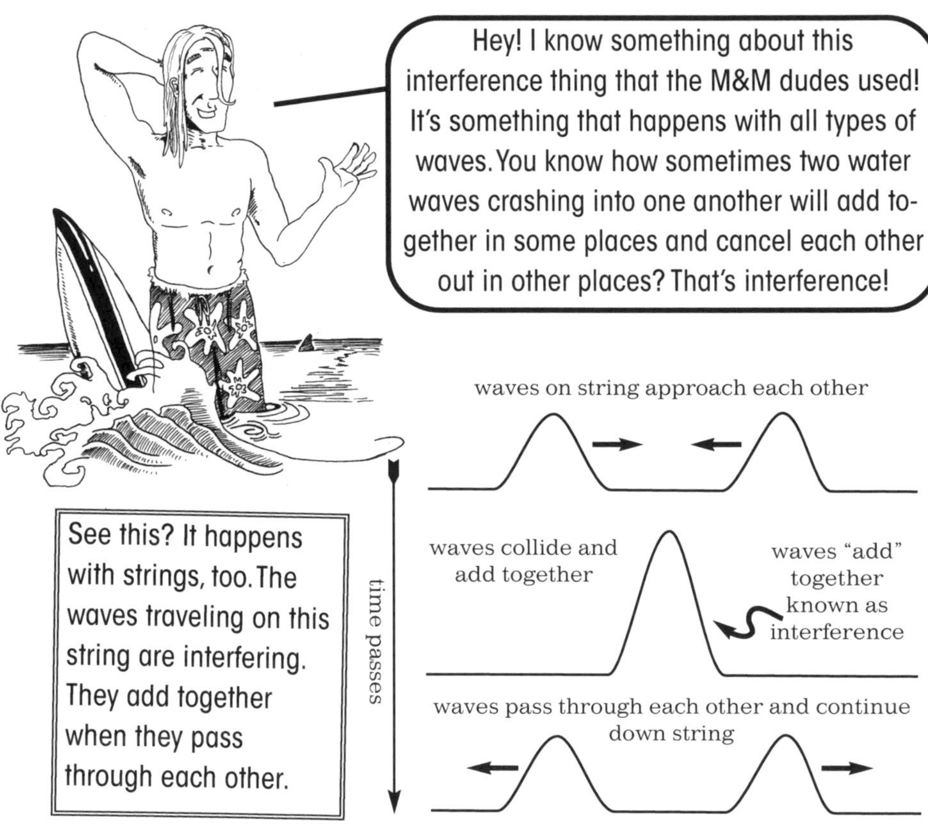
Hey! I know something about this interference thing that the M&M dudes used! It's something that happens with all types of waves. You know how sometimes two water waves crashing into one another will add together in some places and cancel each other out in other places? That's interference!
See this? It happens with strings, too. The waves traveling on this string are interfering. They add together when they pass through each other.
time passes
waves on string approach each other
waves collide and add together
waves "add" together known as interference
waves pass through each other and continue down string

Albert Michelson was the first American awarded the Nobel Prize in Physics. The award was presented to him in 1907.

Wow! Surfer Dude, tear it up. You're right about interference in water waves. Speak quietly lest you trash your rep.

Oh. Yeah. Right. I didn't think about that. Can we get rid of that last page?

Nature has many examples of waves. They all operate more or less like the "wave" that crowds do in big sporting events. In the human wave in a stadium, each person moves up and down with a very particular timing. The wavelike form that travels around the stadium comes about because each person moves a moment after the person to one side and a moment before the person to the other side. This organized and carefully timed movement of many individuals leads to something more than incoherent individual movements. The end result is the wave shape moving around the stadium. In water waves, the water molecules move up and down. In sound waves in air, the air molecules wiggle to and fro. In waves traveling on a guitar string, the string vibrates up and down.

ATOMIC POP
SCIENCE WEEKLY
QUANTUM AIRLINES

Light is also a wave—though it is a bit strange in that there is no waving material. In light, electric and magnetic fields do the waving. More on this later.

But, I want to know about it NOW!
Cooleth thy jets, seeker of knowledge. All will come in good time. Remember, the meaning of life is pudding.

Scientists are humans. They don't like give up their long-held beliefs any more easily than anyone else. For years after the results of the Michelson-Morley experiment were known, physicists struggled to reconcile the experiment with the intuitive vision of relative velocities.

Duh! Maybe the experiment is right after all. You know? I think I can see the forest in spite of the trees. Listen to this . . .
I have no special talents. I am only passionately curious.
—Albert Einstein, in a letter to Carl Seelig

Brilliant and influential people often have the skill of being able to see things and ask questions that seem obvious and simple afterward. Albert Einstein was particularly gifted in this way.

Let's play pretend! Let's make a couple of assumptions and see what pops out.
special relativity
For those who do not think, it is best at least to rearrange their prejudices once in a while.
—Luther Burbank

Einstein created the **SPECIAL THEORY OF RELATIVITY** when he pondered how to relate the observations of two people moving at a constant velocity with respect to one another. He made *two* assumptions that underlie the theory:

Assumption 1

The speed of light is the same for everyone no matter how fast they move.

Hey, the speed of that light is 670,616,630,000 miles per hour!
RADAR
The "special" in the special theory of relativity means that the theory is valid only for situations where the observers move at a constant velocity with respect to each other. If the velocity of one or both of them is changing, the special theory of relativity is not applicable. Einstein invented the general theory of relativity to handle those cases.
Always moving at one speed? How Incredibly boring!
"Relate the observations" of two people? What's that supposed to mean?

Any event—a leaf falling, an egg breaking, a star forming—can be observed by any number of people, even those who are moving with respect to one another. Relativity relates what one person sees to what the other people see when they all look at the same event. "Shouldn't they all see the same thing?" you ask. Most of us would think so. Read on . . .

Assumption 2
The laws of physics are the same for everyone no matter how fast they move.
Hey, if I drop this, it falls.
Yep, it sure does.
That's it?! Dude, I could have thought of that.
ΦBK
Yes. That's it. With these two simple assumptions relativity was born. Let's see where these assumptions lead ...
Let's do a thought experiment starting with these two assumptions. Now suppose we have two observers, Sam and Sally ...
Why Sam and Sally? I hate the name Sam. Why not Biff and Buffy?
Sigh. Okay. Suppose we have two different observers, Biff and Buffy. Biff is riding in the back of a big eighteen-wheeler tractor trailer while Buffy is watching the truck pass from the side of the road.
PHYSICAL MOVERS
Get in Buffy. Let's ride.
No way. I'll watch from here. My mom told me never to get in the back of trucks with boys.

Okay, Biff is in the trailer of the truck, and Buffy is sitting on the side of the road. The truck moves at a constant velocity and zooms past Buffy. Imagine that a special lamp sends a pulse of light from the truck trailer floor to the ceiling where it is reflected off a mirror back down to the floor. Both Biff and Buffy observe this same event. (Buffy can see it because the side of the truck is made of glass.) Further, let us suppose that the inside of the truck is sort of foggy so that both Biff and Buffy can see the path that the light takes as it moves from the floor to the ceiling and back down to the floor of the truck. What do Biff and Buffy see, and how do we relate those observations?

Excuse me. I'm Thad, one of Biff's fraternity brothers. I'm here for the show. Biff and Buffy each have to see the same thing, right? They are looking at the same thing after all.
ΦBK

They are looking at the same event. True. But, how do each of them perceive it?

Biff is moving with the truck. So he perceives the light to move straight up to the ceiling and straight back down to the floor.

Buffy, on the other hand, watches the truck zoom by as the light travels to the ceiling and back. Because light travels at a finite speed, she sees the truck move forward during the time that the light is traveling. So she perceives the light to move in something of a triangular path.

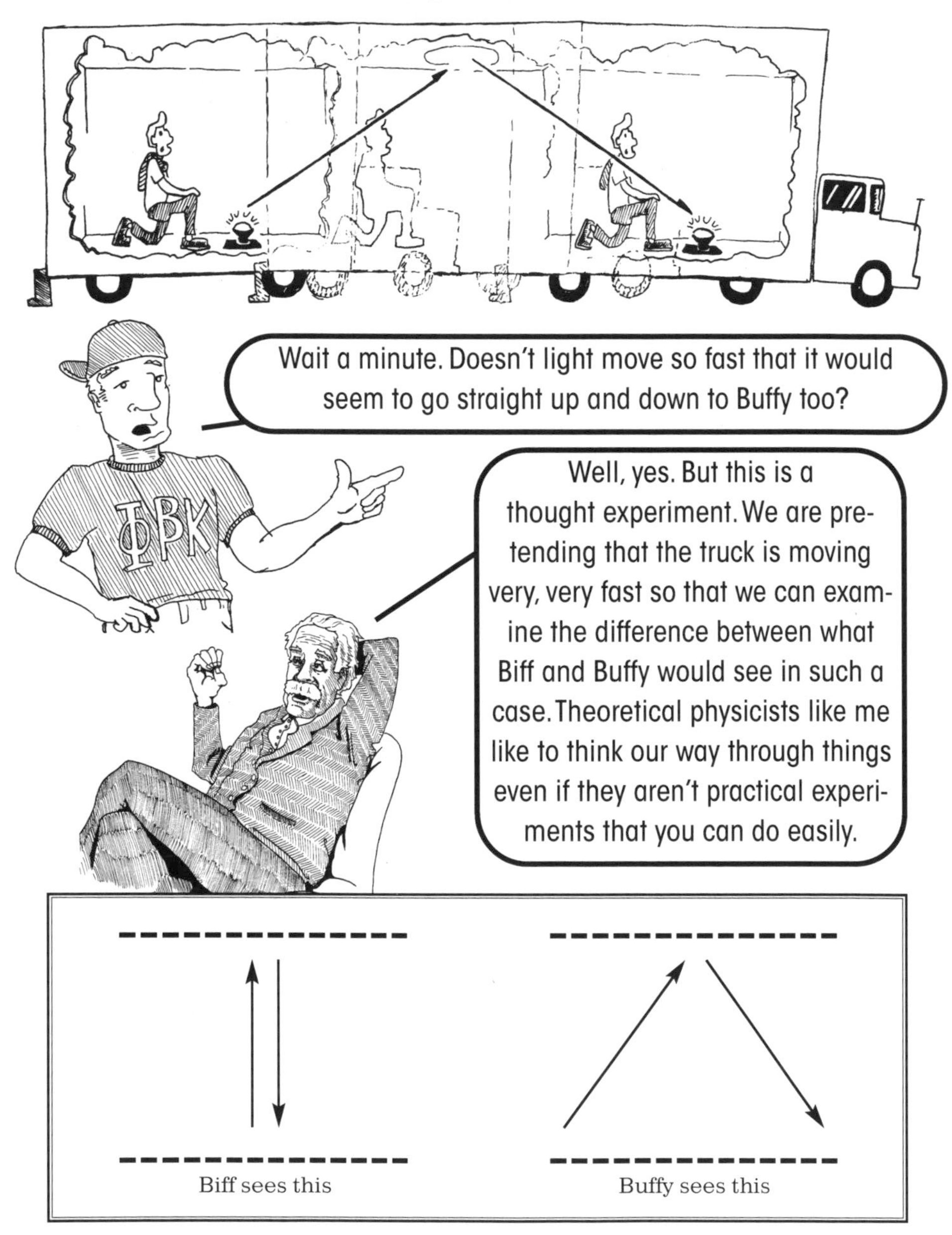
Wait a minute. Doesn't light move so fast that it would seem to go straight up and down to Buffy too?
Well, yes. But this is a thought experiment. We are pretending that the truck is moving very, very fast so that we can examine the difference between what Biff and Buffy would see in such a case. Theoretical physicists like me like to think our way through things even if they aren't practical experiments that you can do easily.
Biff sees this
Buffy sees this

By virtue of their different points of view, Biff and Buffy see the light path differently. Buffy perceives the light to travel further than Biff.

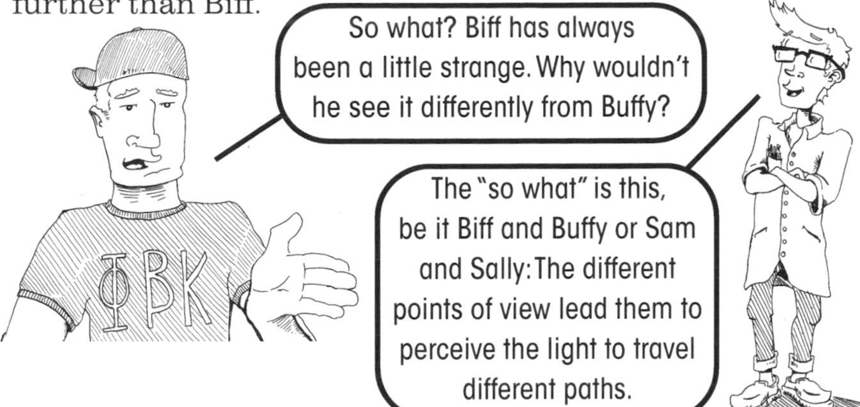
So what? Biff has always been a little strange. Why wouldn't he see it differently from Buffy?
The "so what" is this, be it Biff and Buffy or Sam and Sally: The different points of view lead them to perceive the light to travel different paths.

The speed of light is the distance the light travels divided by the amount of time it takes to travel that distance. Buffy sees the light travel a longer distance. Since Einstein's first assumption is that the speed of the light is the same for Buffy and for Biff, it means that the time it takes the light to travel to the ceiling and back is longer for Buffy than for Biff.

Excuse me?
Time is relative and not absolute.
What Al is trying to say is that if we are to assume light travels at the same speed as seen by two observers moving with respect to one another, it means that time must pass at different rates for the two observers! In other words, if Buffy perceives the light to travel farther than Biff and the speed of light is the same for both of them, the event must take more time for Buffy. Time moves faster for Buffy than for Biff.
Oh boy. I knew I should have stayed in bed this morning.

Let's go back to the example where we handed out synchronized watches to ten different people in a room and asked them to leave, go about their business, and return in exactly one hour. Our typical, intuitive, human view of time and space tells us that each of the people would return to the room at the same time and with the same time reading on each of their watches, regardless of what they did during that hour. Einstein tells us that's not the case. Relative to one another, the time on their watches would depend on how fast they moved during that hour. Time is *relative*.

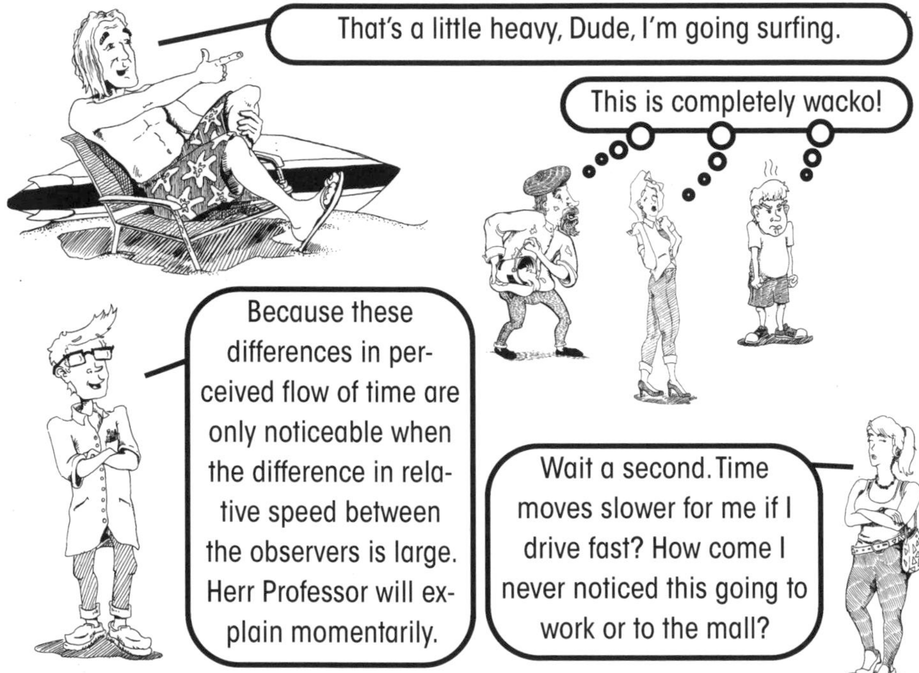
That's a little heavy, Dude, I'm going surfing.
This is completely wacko!
Because these differences in perceived flow of time are only noticeable when the difference in relative speed between the observers is large. Herr Professor will explain momentarily.
Wait a second. Time moves slower for me if I drive fast? How come I never noticed this going to work or to the mall?

This stuff may sound wacko, but many, many scientific experiments support the conclusions of relativity. For example, in 1971, J.C. Hafele and Richard Keating compared the flow of time on extremely precise clocks flown by airplanes around the world in each direction with clocks left on the ground. They found the clocks moved at different rates in a way that agreed exactly with relativity.

> Relativity teaches us the connection between the different descriptions of one and the same reality.
> —Albert Einstein

Chapter 4
Herr Professor Lets the Nerds Loose

The effect of motion on time is very small unless the difference in the relative velocity between the two observers is very, very large. Close to the speed of light. Allow me to derive for you the time dilation formula that describes this effect. Then all will be clear.

TIME OUT!

Herr Professor, the publisher wants me to remind you that the title of this book is *Relativity and Quantum Physics For Beginners*. If you start putting in mathematical derivations, they'll have to change the name of the book, and they don't want to do that. No derivations.

Sigh . . . But, it's not so hard, and some readers will have fun following along. Hmmm. Okay, I won't do it (wink, wink).

If you're the publisher of this book or if you don't like math, please skip the rest of this chapter! Otherwise, enjoy this derivation of Einstein's time dilation formula.

The point of this derivation is to determine the relationship between the times that Buffy and Biff measure as the light travels up to the ceiling and back down to the floor. Though it is done in the context of a particular example, the result shows how time passes at different rates for any two observers moving with respect to one another.

h
Biff sees this

If the height of the truck is h, from Biff's point of view the light travels a distance of $2h$ in a time T_{Biff}. Since speed is distance divided by time, the speed of light as seen by Biff is

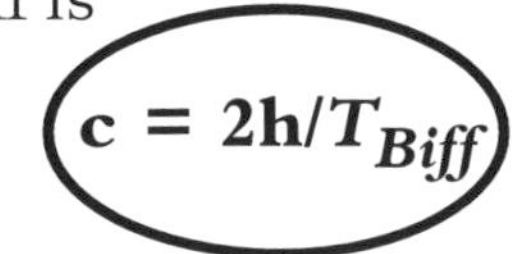

$$c = 2h/T_{Biff}$$

> Mathematics is the door and the key to sciences.
> —Roger Bacon

h
d
d
vTBuffy
Buffy sees this

From Buffy's point of view the light travels from the floor to the ceiling and back to the floor in a time T_{Buffy}. During that time the truck, which is moving at speed v, travels a distance of vT_{Buffy}.

velocity = distance ÷ time, so distance = velocity x time

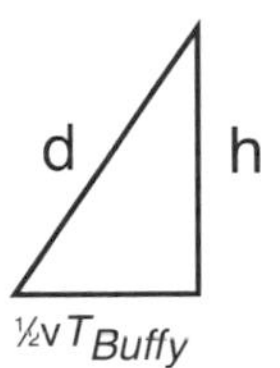
d
h
½vTBuffy

Looking at the diagram above we can see that as the light travels to the ceiling, the distance it travels is one side of a triangle, as shown to the left. Using the Pythagorian theorem that relates the lengths of the sides of triangles, we get

$$\mathbf{d^2 = h^2 + (\tfrac{1}{2}vT_{Buffy})^2}$$

So Buffy sees the speed of the light to be given by

$$\mathbf{c = \frac{2d}{T_{Buffy}} = \frac{2\sqrt{H^2 + (\tfrac{1}{2}vT_{Buffy})^2}}{T_{Buffy}}}$$

Whew! Now comes the fun part. You have one expression for the speed of light in terms of the time Biff measures for the event to happen, and you have one expression for the speed of light that involves the time that Buffy measures for the event to happen. Einstein says these two speeds of light are the same. So, we set them equal to each other and solve for Buffy's time in terms of Biff's time.

Oh . . . I think I'm gonna be sick.

c according to Biff = c according to Buffy

$$\frac{2h}{T_{Biff}} = \frac{2\sqrt{h^2 + (\frac{1}{2}vT_{Buffy})^2}}{T_{Buffy}}$$

Square both sides
$$\left(\frac{2h}{T_{Biff}}\right)^2 = \left(\frac{2h}{T_{Buffy}}\right)^2 + \left(\frac{2}{T_{Buffy}}\right)^2 (\tfrac{1}{2}vT_{Buffy})^2$$

Divide both sides by $(2h)^2$
$$\left(\frac{1}{T_{Biff}}\right)^2 = \left(\frac{1}{T_{Buffy}}\right)^2 + \frac{v^2}{(2h)^2}$$

Use the fact that $2h = cT_{Biff}$
$$\left(\frac{1}{T_{Biff}}\right)^2 = \left(\frac{1}{T_{Buffy}}\right)^2 + \frac{v^2}{(cT_{Biff})^2}$$

Multiply both sides by T_{Biff}^2
$$1 = \left(\frac{T_{Biff}}{T_{Buffy}}\right)^2 + \left(\frac{v}{c}\right)^2$$

Here is the bottom line.

Rearrange to solve for T_{Buffy} in terms of T_{Biff}
$$T_{Buffy} = T_{Biff}\left(\frac{1}{\sqrt{1-(v/c)^2}}\right)$$

Look! I can chew gum and ride a bike!

According to Einstein's special theory of relativity, the relationship between the time measured in one frame of reference—or point of view, such as Buffy's point of view—and another frame of reference, such as Biff's point of view, moving at velocity v with respect to the first is

$$T_{Buffy} = T_{Biff}\left(\frac{1}{\sqrt{1-(v/c)^2}}\right)$$

This term is 1 if v is much smaller than the speed of light. In this case $T_{Buffy}=T_{Biff}$. Since everyday life consist of relative motions that are much smaller than the speed of light, this is what we experience daily and what drives our intuition. Time seems absolute—the same for everyone—because the "relativistic" effect is too small to notice.

Physicists usually call this quantity γ, which is the Greek letter gamma.

When the relative velocity between the two observers (v) is close to the speed of light, this term becomes large and T_{Buffy} is bigger than T_{Biff}. In this case, the passage of time is perceived to be very different, depending on your point of view. For example, if Biff and the truck move past Buffy at 98 percent of the speed of light, this term is 5. In that case, 5 seconds would pass for Buffy for each second that passes for Biff.

The meaning of life is STILL pudding, even if time is relative.

Dude, like clue me in on something practical. Suppose I want my date to last longer this Friday night. Do I drive my car faster or slower?

Chapter 5
Relatively Speaking

Say what?! You mean we aren't special??

Nothing said so far about relativity is specific to our friends Biff and Buffy.

Sorry Biff. Sorry Buffy. But you hadn't even been born when Al discovered this stuff in 1905. The deal is this: If something happens that takes a certain amount of time as measured by someone standing still with respect to that event, the time for that event to is occur is longer for anyone moving with respect to the event. The difference is big if their speed relative to the event is very large—say, close to the speed of light. The difference is not noticeable at all if the relative speed is small—say, like the speed of a bus traveling down Main Street.

For example, suppose a snake eats a mouse in one minute according to Buffy, who happens to be standing still by the snake watching the demise of the poor mouse.

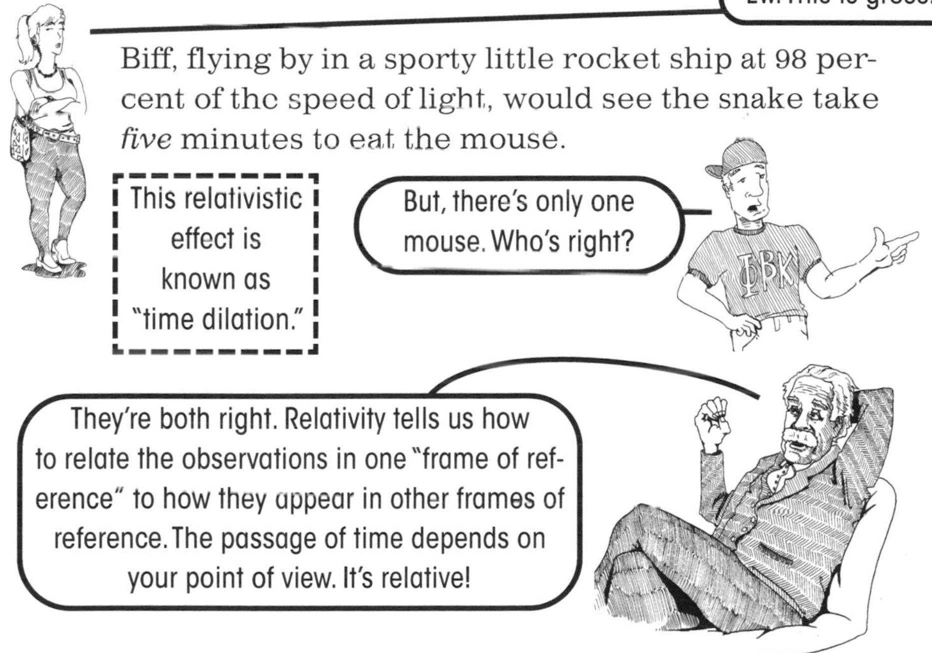
Ew. This is gross.

Biff, flying by in a sporty little rocket ship at 98 percent of the speed of light, would see the snake take *five* minutes to eat the mouse.

This relativistic effect is known as "time dilation."
But, there's only one mouse. Who's right?
They're both right. Relativity tells us how to relate the observations in one "frame of reference" to how they appear in other frames of reference. The passage of time depends on your point of view. It's relative!

Relativity does not deal specifically with *time*. Relativity deals with *anything* that you might observe, or measure, from different points of view, including time.
Like what kinds of things? Girls? Man, I like looking at girls from different points of view.
Sigh. Surfer Dude, this is a family publication, okay? We're talking more about things a scientist might measure, like time and space (or positions), and energy and forces.

From Einstein's two assumptions and an experiment similar to what we saw with Biff and Buffy and the truck, it can be shown that space is not absolute any more than time is absolute. Our intuition of space as three immutable dimensions sliding through time is incorrect. Relativity tells us that one's perception of position or space depends on your frame of reference in a way that is eerily similar to what we saw with time.

To see what relativity says about space, let's return to our snake eating a mouse. Buffy, who is sitting still by the snake, whips out a meter stick and finds that the snake is one meter long.

This is still gross.

Biff, on the other hand, looks at the snake as he zooms by at 98 percent of the speed of light, and he measures the snake to be only 20 centimeters long, or a factor of 5 shorter. Again, the strange length contraction is only noticeable if Biff zooms past at a large fraction of the speed of light.

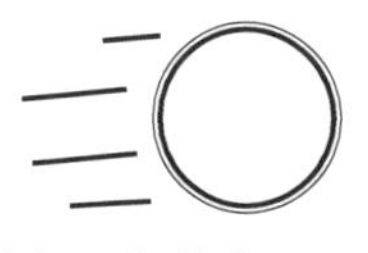

A baseball thrown past you by a good pitcher looks like this.

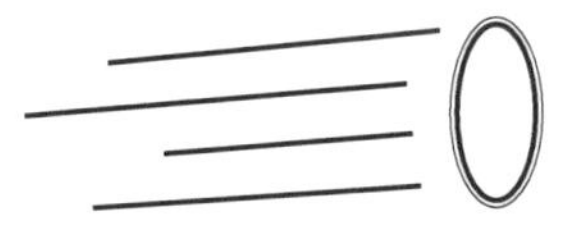

A baseball passing you at a huge speed—say, 98 percent of the speed of light—would look more like this.

This relativistic effect is known as "length contraction."

If relativistic length contraction and time dilation aren't strange enough for you, get a load of this . . .

So far, we have discussed space and time as independent things. But, I was a stickler for doing things right. Take it away Herr Professor . . .

Thanks Al. Professor Einstein discovered how to relate the observations in one reference frame (or point of view) to the observations in another reference frame. But time and space aren't really separate. In general, in every reference frame, each event happens at a specific time and place. It turns out that the position in one frame depends on both the time and position in the other frame. Also, the time in one frame depends on both the time and position in the other frame. Physicists call the mathematical formulas that relate the observations between reference frames "Lorentz transformations."

If you like math, what Herr Professor is trying to say is this:

Suppose Buffy sits by the side of the road and sees a firecracker explode at time t and position x (along the road). If Biff zips by in a fast car at speed v and passes Buffy right at $t=0$, relativity tells us that Biff would see the firecracker explode at a position x' and a time t' given by

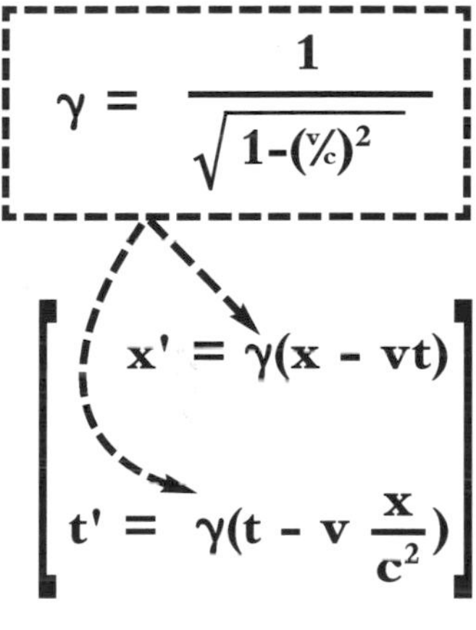
$\gamma = \frac{1}{\sqrt{1-(v/c)^2}}$
$x' = \gamma(x - vt)$
$t' = \gamma(t - v\frac{x}{c^2})$

Note that the position and time get all mixed up under the "transformation" from one point of view to the other. If v is small—meaning Biff is going slow—γ is equal to 1 and $x'=x$ and $t'=t$. In other words, Biff and Buffy see the same thing. If v is large—like a big fraction of the speed of light (say 0.9c)—γ is large, and Biff and Buffy see very different things.

Because time and space get all mixed up as you relate measurements from one point of view to another in relativity, physicists no longer think of space and time as separate things. Instead, physicists talk about *space-time*.

Besides *space-time* sounds really cool!

Why do physicists call the equations that relate observations between different frames of reference "Lorentz transformations"? Weren't they invented by Albert Einstein? Why not call them "Einstein transformations" or "Al's fabulous freaky formulas"?

Einstein wasn't the only person to have thought deeply about these matters. In fact, the Lorentz transformations were known before Einstein's work in 1905. Hendrik Lorentz and others had uncovered these relationships during earlier investigations, and many of their ideas are contained in Einstein's special theory of relativity.

Lorentz was a Dutch physicist who received the 1902 Nobel Prize in Physics for his work aimed at understanding the effect of magnetism on light emitted by atoms.

It seemeth to me that pudding is absolute and not relative. This I shall calleth the "principle of relativistic pudding invariance."

> Whatever came from this supreme mind was as lucid and beautiful as a good work of art. He meant more to me personally than anybody else I have met in my lifetime.
>
> —Albert Einstein, in an essay written to celebrate the hundredth anniversary of Lorentz's birth

Relativity is what we use to relate observations from one point of view to another.

Any observation? What if I observe that my friend Buffy is in a bad mood? Would someone moving with respect to me need relativity to sense Buffy's mood?

Oh, no! Not that kind of observation. I'm talking about things that scientists measure, like position, time, energy, mass, momentum, force, and so forth. Good luck, by the way. She looks pissed.

Let's give them an example, Al. One of the fundamental principles of physics as we know it is called *momentum conservation*. Momentum is mass x velocity. When two things collide, the speeds and directions of motion of the objects change, but the total value of momentum stays the same (or is *conserved*, as a physicist would say). This is why big, fast-moving football players keep moving forward when hit by smaller players. Because momentum and momentum conservation are so very important and useful in physics, it is important that we understand how momentum appears as measured from different points of view. We need to see how momentum transforms from one reference frame to one that is moving with respect to the first.

Also, remember that one of the basic assumptions underlying relativity is that the laws of physics are the same for everyone no matter how they move. This means that momentum must be conserved according to all observers, no matter how they move.

Ugh!

Grrr! I have momentum on my side!

Just as space and time get all mixed up when observations are related across reference frames, Einstein discovered that in order for momentum conservation to hold true for all points of view (as it must), momentum also gets mixed up with something else. That something else is known as *relativistic energy, E.*

$$\mathbf{E = \gamma mc^2 = \frac{mc^2}{\sqrt{1-(v/c)^2}} = \underbrace{mc^2}_{\text{term 1}} + \underbrace{\tfrac{1}{2}\, mv^2}_{\text{term 2}} +} \text{ terms that are so small that they can be ignored unless v is close to the speed of light}$$

The *second term* above is known as *kinetic energy*. This is the energy of motion, and physicists understood this as a form of energy long, long before Einstein. But notice that in the case where v=0, that is, you are observing a mass at rest, the *first term* is still present. Relativity says any object with mass has a *rest energy*. The idea that mass and energy are different faces of the same thing was not new with Einstein, but the fundamental nature of the relationship and its relation to space and time transformations was another jolting revelation brought to light by the theory of relativity!

E=mc²
Hey man, tell me something . . . I've seen this equation all my life. What's the deal? Why should I give a crap about it? Will it give me better waves?
Better waves? Probably not.

This equation is incredibly important for fundamental science and to society as a whole. It also holds the key to understanding the forces of nature, stars, and the evolution of the universe.

If you want to do anything, it requires energy. One of the fundamental laws of physics is known as the law of conservation of energy. What it means is that—in spite of how it might seem—energy cannot be created or destroyed.

Yeah . . . right. Come on, man. We started with a cold pile of wood and now we have this toasty fire. You're telling me that we didn't create energy? Sure we did.
No. You didn't *create* energy. You just *changed its form.* That's allowed under the laws of physics.

Energy takes many different forms. To see this, it's good to start with thinking about what you mean when you say *energy*. Informally, a physicist thinks of energy as the ability to move things. The most obvious form of energy is the energy of motion. A moving car has the ability to move things, because if it runs into something, that something will move. A stretched spring has energy, because if it is attached to something and released, it can move that thing. The energy in that stretched spring is called *potential energy* because it's stored until the spring is released—it has the *potential* to move things.

A bowling ball held high has (gravitational potential) energy because you can drop it and move things.

TNT has (chemical) energy because it can explode and move things.

Energy can be in the form of heat, light, sound, water waves and on and on . . .

AND energy can be in the form of mass. Or perhaps it is better to say that energy and mass are different faces of the same thing. Cool physics dudes like me call this idea *energy-mass equivalence*.

When a pot of water is heated on a stove, the water gains heat energy. It is also thought to become slightly more massive, though the change in mass is too small to measure.

Energy-mass equivalence is a critical concept in modern physics. It is important for our understanding of quantum mechanics, forces, and the origin of the universe. We'll talk about much of this later in the book. The fun part is the realization that it is possible to convert mass energy to other forms of energy and vice versa.

Hey, my wife keeps telling me my belly is getting so big that it qualifies for an additional deduction on our taxes. You think you can convert some of this mass to energy? We could call it the Einstein diet? Eh? And do late-night infomercials about it. Whaddaya think?

Uh . . . I don't think so . . . But here's something cool to think about. Matter in the world around us is made of atoms. The atoms are mostly empty space. Each atom consists of an extremely tiny *nucleus* surrounded by even tinier particles called *electrons*. Most of the mass in each atom is in the nucleus, which is made up of particles called *protons* and *neutrons*.

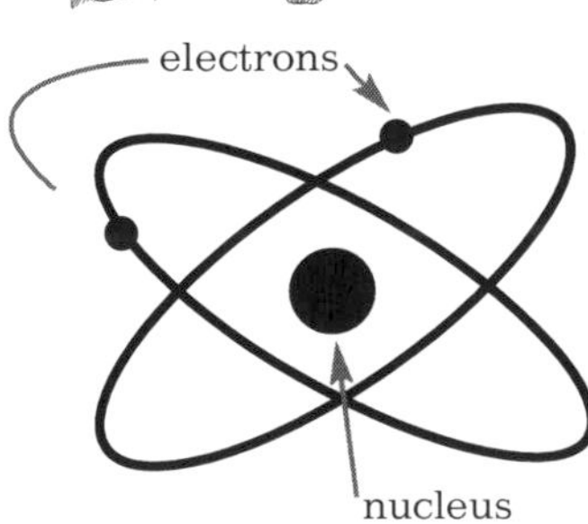
electrons
nucleus

There are slightly more than one hundred different types of atoms in the universe. Each distinct type of atom—known as an element—differs from the other types of atoms by the size of its nucleus or, more specifically, by the number of protons in its nucleus. Hydrogen atoms have one proton, for example, while uranium has 92 and carbon has 6.

So, what is it that happens when Camper Bob starts a fire? How is it that energy gets released in the form of heat and light?

Well, I'll tell ya. The atoms in wood are arranged in structures called molecules. When the wood burns, the large wood molecules break up into smaller molecules, which store less energy. The difference in energy is given off as light and heat. If we could measure incredibly tiny differences in mass, we would see that the mass of the wood is slightly larger than the mass of products formed when the wood burns.

Wow! What happened to my friend Camper Bob? Did aliens abduct him and leave me a rocket scientist in his place?

When atoms in molecules are rearranged into other molecules, the process is called a *chemical reaction*. Sometimes you have to put energy into the system to make the reaction happen. Sometimes you get energy *out* of the system. For example, when a grenade explodes or gasoline burns, you get energy out of the system. A similar thing happens with nuclei. In "nuclear reactions" the protons and neutrons are rearranged into different nuclei. The energy differences involved in nuclear reactions are about a million times larger than the energy differences involved in chemical reactions. That's why normal bombs go *boom* and nuclear bombs go *KA-BOOM*!

If a very large atomic nucleus splits into two smaller nuclei, The process is called *nuclear fission*. This is the process that drives nuclear reactors and nuclear bombs.

If two small atomic nuclei are fused together, the process is called *nuclear fusion*. Nuclear fusion is used in the largest nuclear warheads. More importantly, nuclear fusion is the primary power source that drives the sun and other stars.

Oh! Those kids are at it again. Dang! I told them fireworks aren't allowed at this campsite!
Uh . . . that looks a little bigger than a firecracker, Bob.

Hey man, You got any marshmallows in this strange RV?
No campfires in here, Earthling.

In nuclear war, all men are cremated equal.
—Dexter Gordon

The discovery of nuclear chain reactions need not bring about the destruction of mankind any more than did the discovery of matches. We must only do everything in our power to safeguard against its abuse.
—Albert Einstein

Nature is neutral. Man has wrested from nature the power to make the world a desert or to make the deserts bloom. There is no evil in the atom; only in men's souls.
—Adlai Stevenson

Chapter 6
Surfing the Warped Fabric of Space-Time

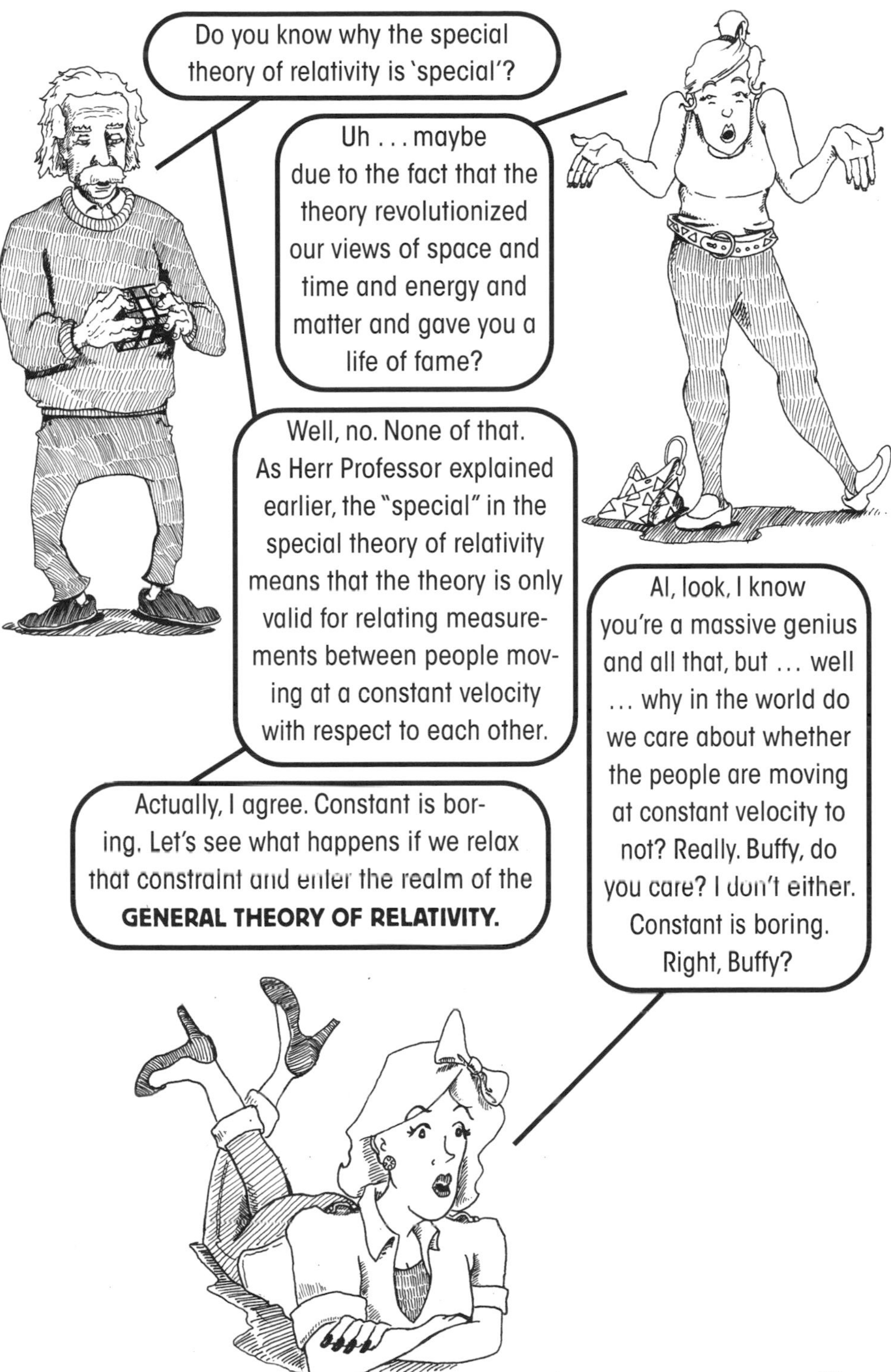
Do you know why the special theory of relativity is 'special'?
Uh . . . maybe due to the fact that the theory revolutionized our views of space and time and energy and matter and gave you a life of fame?
Well, no. None of that. As Herr Professor explained earlier, the "special" in the special theory of relativity means that the theory is only valid for relating measurements between people moving at a constant velocity with respect to each other.
Al, look, I know you're a massive genius and all that, but ... well ... why in the world do we care about whether the people are moving at constant velocity to not? Really. Buffy, do you care? I don't either. Constant is boring. Right, Buffy?
Actually, I agree. Constant is boring. Let's see what happens if we relax that constraint and enter the realm of the GENERAL THEORY OF RELATIVITY.

Have you ever noticed that you feel as if your weight changes when you are on a rapidly accelerating elevator? If the elevator is accelerating upward you feel heavier. In fact if you were to stand on a set of scales while on the elevator, the scale would say you weigh more.

Oh my goodness! I'm never eating again. Look at that!

In fact, if you are in a closed box—like an elevator—and not allowed to look outside, it is not possible to distinguish whether you are in a little room on earth or whether you are in a box far out in space that is accelerating just enough to press you against the floor with the same force you would feel on earth.

Oh puh-lease! If you want to say something scientific, you need to say it so that it sounds more impressive than that! What I said way back in the day was that it is impossible to distinguish an accelerating reference frame from gravitation. This is known as the *equivalence principle*.

The equivalence principle means that the general theory of relativity not only describes how to relate accelerating reference frames, but it is also a *theory of gravitation*.

By giving up on the demand that things move at constant velocity with respect to each other, special relativity no longer works well. In the *general theory of relativity*, Einstein studied what happens when you relate measurements between points of view that accelerate with respect to each other.

So this Einstein guy uncovered a pretty strange world with his special theory of relativity. I almost hate to ask . . . what kind of weirdness happens in the general theory of relativity?

Are you sure you want to know? If so, follow me into into this little room . . .

Gravity feels like acceleration. When you are in a car and the driver steps on the gas, you are pressed back into the seat as the car surges forward (or accelerates). This is the essence of Einstein's equivalence principle.

If you are in a little room with no windows, it is not possible to distinguish the two situations sketched below.

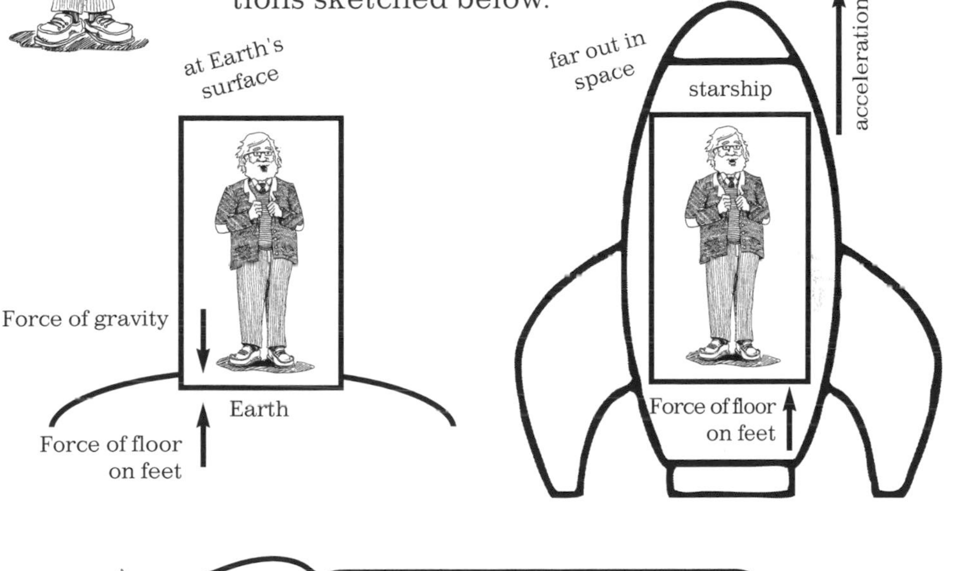

at Earth's surface
Force of gravity
Earth
Force of floor on feet
far out in space
starship
acceleration
Force of floor on feet

A physicist says that something *accelerates* if its velocity changes. If it is moving in a straight line that means the object is speeding up or slowing down.

Imagine that Biff is out in space in a little room on a rocketship that is accelerating—moving faster and faster—and pressing Biff against the floor in such a way that he can't really tell whether he is on a rocketship or standing on Earth. Buffy floats out in space near the path of Biff's rocket. She watches it pass by. As the rocket passes by Buffy, Biff pulls out a little ray-gun and shoots a short pulse of light toward the wall. If Biff and Buffy could see the path of the light pulse, what would they see?

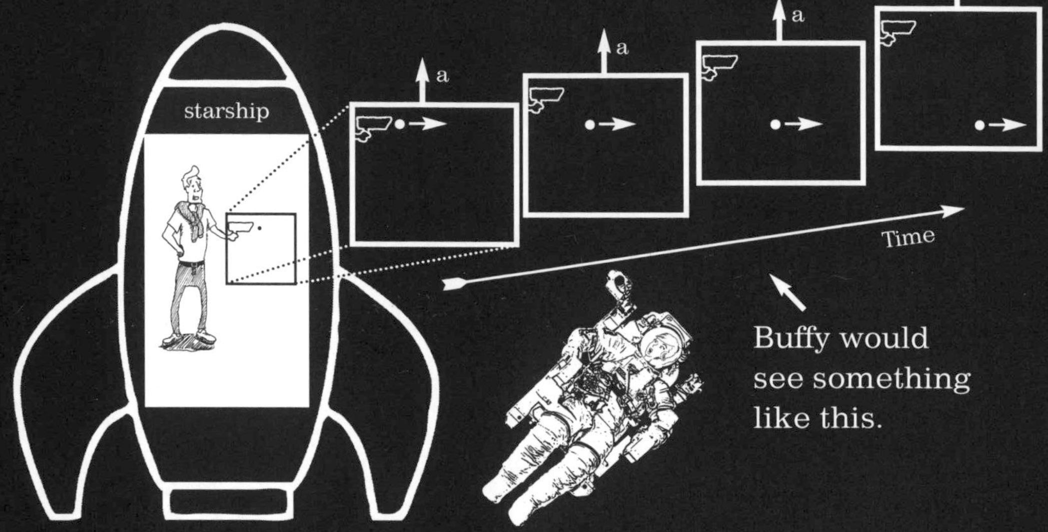
a
a
a
a
starship
Time
Buffy would see something like this.

To Buffy, the light would appear to move in a straight line toward the wall while the room (attached to the rocket) moves faster and faster (in the direction denoted by the arrow labeled 'a'). Successive snapshots might look something like what is sketched above. This little game assumes that Buffy can see through the wall of the rocket and can see the light pulse. All of that is crap, of course. But, that's okay. You can still pretend it's true and see what you learn in the process.

Biff, on the other hand, would see the pulse of light drop toward what he thinks of as the floor of his room since the rocket would move upward faster and faster as the pulse of light moves toward the wall. Biff would see something like this.

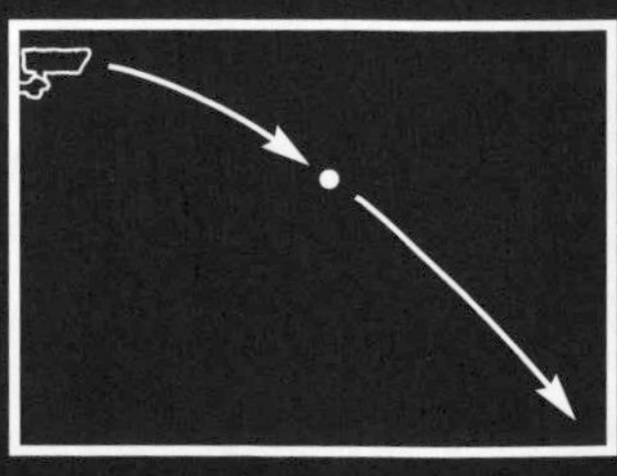

Oh-oh! I sense weirdness. Here it comes! Take it away, Herr Professor.

Biff, moving in an accelerated reference frame—er, I mean a rocket—sees the pulse of light move along a curved path. Einstein's principle of equivalence says that we can't tell the difference between an accelerated reference frame and gravity. This must mean that gravity causes light to travel along a curved path. But light travels along the shortest distance between points. This led Einstein to conclude that gravity bends space! Light travels along a curve because space itself is bent.

Of course, we already know that space and time are all mixed up. So we can't really speak only of space. Einstein's general theory of relativity is a theory of gravitation built on the idea that gravity arises from the curvature of space-time. It's a geometric theory using geometric rules that are a little different from what we typically learn in middle school.

In the special theory of relativity, we saw that space and time are relative, not absolute. We also saw that space and time are intimately connected as one—space-time. In the general theory of relativity, space-time is viewed as a fabric that can be warped. Those distortions are what we perceive as gravity. Also, the presence of gravity means there is a distortion in space-time.

I don't know about this. Are you sure all you physicists didn't just make this crazy stuff up?

What do you mean by a "distortion of space-time"? Do you mean like how the last two minutes of a football game last forever? Or maybe how cheesecake seems to distort thighs?

A few of the observations supporting relativity

Einstein's general theory of relativity is a scientific theory, which means it's not all that useful if the conclusions of the theory do not agree with experimental observations. What is nature's verdict?

Us science types have found that general relativity is difficult to test because, under normal conditions, the theory predicts only very tiny differences from what we would normally expect to see using Newton's theory of gravity. Still, we have managed to confirm that general relativity appears to describe the universe quite well.

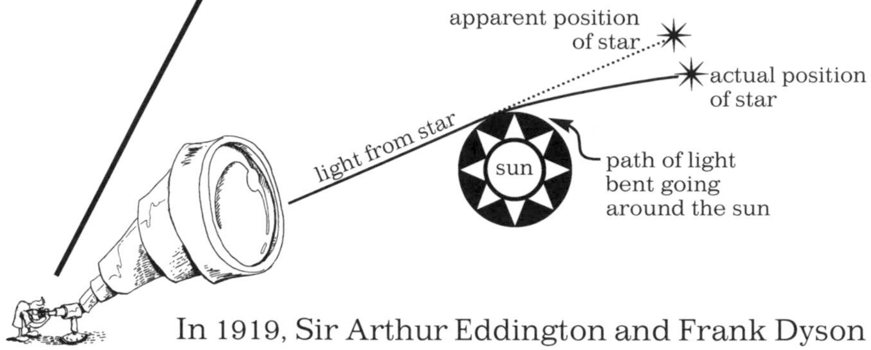

apparent position of star

actual position of star

light from star

sun

path of light bent going around the sun

In 1919, Sir Arthur Eddington and Frank Dyson led a team that observed the bending of light in space near the sun. They inferred the bending of space by measuring how the positions of distant stars changed when the light passed close to the sun during a solar eclipse. The bending of light around the sun agreed with the amount predicted by general relativity.

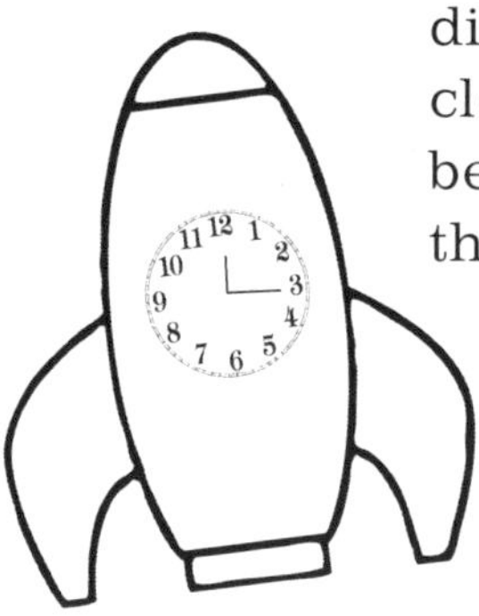

General relativity predicts that time passes more slowly in places where there is a strong gravitational force. Scientists have sent very accurate clocks high up in rockets where the force of gravity is weaker than on Earth and verified that clocks run more slowly near the surface of the Earth.

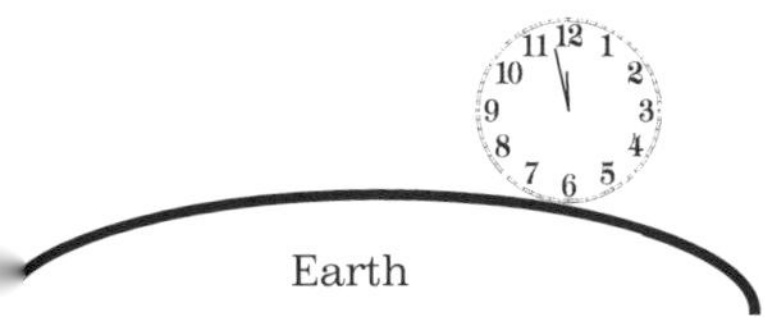

Earth

The closest point of approach of the planet Mercury to the sun on each orbit rotates about the sun in a way that is predicted by general relativity. This motion is not predicted correctly by Newton's theory of gravity.

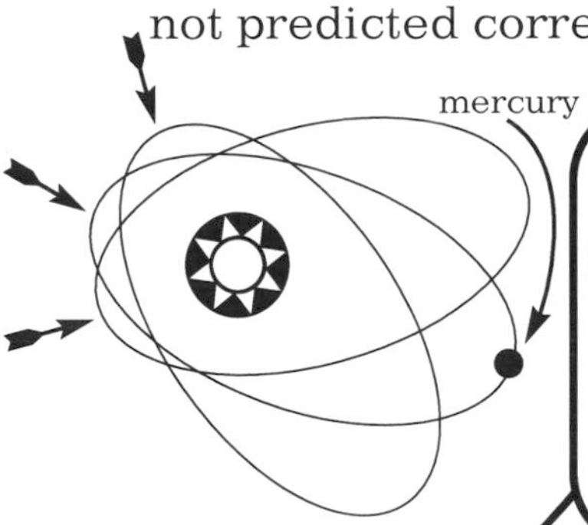

mercury

Point of closest approach of Mercury to the sun moves around the sun on successive orbits

You know what? I manage to eat my oatmeal and go to school without ever having to worry about taking strange routes or resetting my clocks because of the current day's distortions of space-time. And I've never even seen the planet Mercury. So WHY DO I CARE about this??

Well, here's one reason: The Global Positioning System, or GPS as it is commonly known, consists of a series of satellites orbiting the earth. GPS receivers decode signals from these satellites and can determine the position of the receiver to approximately 10 inches. This amazing accuracy is achieved only with the inclusion of relativistic effects in the calculations. In case you haven't thought about it, the GPS system can do much more than simply keep you from getting lost on the way to a new friend's house. The GPS system is invaluable for navigation and shipping, with both commercial and military uses.

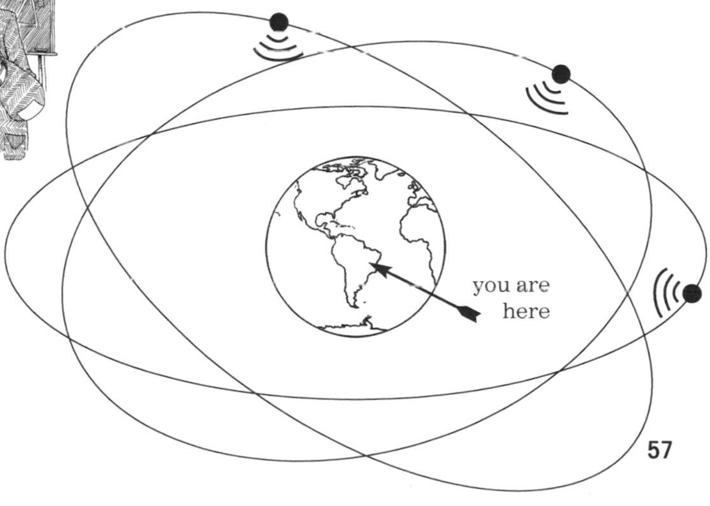

you are here

General relativity predicts some strange things. Perhaps the strangest of these things is called a **BLACK HOLE.**

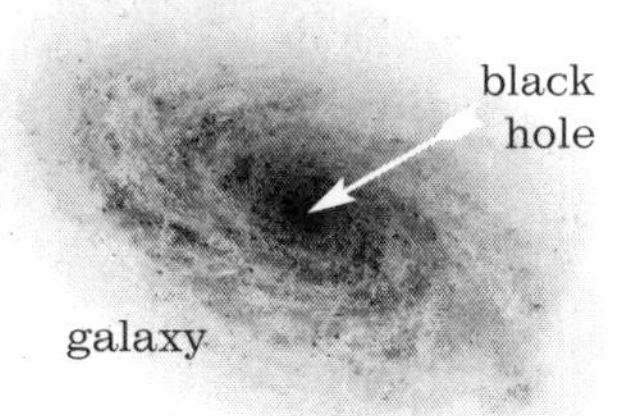
black hole
galaxy

A black hole is an object that has a very strong gravitational force. Space is bent so much near a black hole that light can't escape. The distortion of space-time is so strong that to an outside observer time appears to stop at the edge of the black hole. Because light does not escape this object, it looks like—well—a black hole. We can only detect black holes indirectly by measuring the gravitational effect of the black hole on nearby objects or the light emitted by matter as it is ripped apart as it spirals into the black hole. Black holes are created in the final stages of life for very massive stars. Also, black holes are thought to exist at the center of large galaxies.

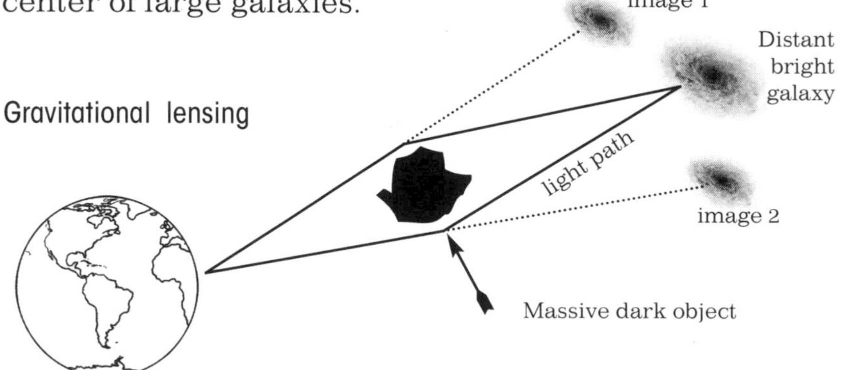
Gravitational lensing
image 1
Distant bright galaxy
light path
image 2
Massive dark object

Light from very distant objects in the universe is sometimes bent by masses along the path of the light. This phenomenon is called *gravitational lensing*. Astronomers can use the image of the "lensed" object to map out the distribution of mass in the universe.

Dude! If I had a little black hole stuck to the back of my board, I'll bet it'd bend the water into some AWESOME waves!
Where did you find this Neanderthal?

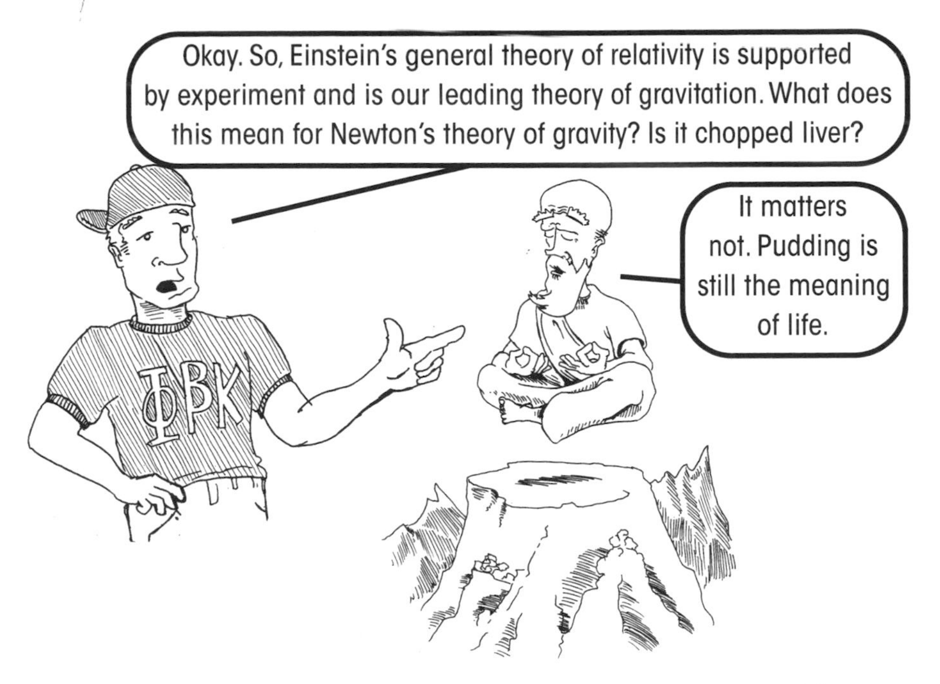
Okay. So, Einstein's general theory of relativity is supported by experiment and is our leading theory of gravitation. What does this mean for Newton's theory of gravity? Is it chopped liver?
It matters not. Pudding is still the meaning of life.
ΦBK

To date, all scientific observations are consistent with general relativity. Some observations—such as the orbit of Mercury—cannot be accounted for by Newton's theory of gravity. So, scientifically, Einstein's theory is the winner. But does that mean we throw out Newton's theory? No. Newton's theory works very well under conditions normally encountered on Earth. In fact, under everyday circumstances the equations of general relativity are the same as Newton's equations. Engineers can construct buildings and bridges that stand for decades without knowing anything about general relativity. In science, it is fair to use the simplest theory that is applicable—remember Ockham's razor.

> Theories should be as simple as possible, but no simpler.
> —Source unknown, often attributed to Albert Einstein

> Science is a cemetery of dead ideas.
> —Miguel de Unamuno

> Science is always wrong. It never solves a problem without creating ten more.
> —George Bernard Shaw

Chapter 7
Relativity Light

Dude! What's going on? I can't read this!
Duh! Hey moron, I grew up in Germany and published this paper in a prestigious German scientific journal called *Annalen der Physik*. So, of course I wrote in it German. Sprechst du Deutshe?
Behold the first page of Einstein's original paper on special relativity published in 1905.

3. *Zur Elektrodynamik bewegter Körper;*
von A. Einstein.

Daß die Elektrodynamik Maxwells — wie dieselbe gegen-
wärtig aufgefaßt zu werden pflegt — in ihrer Anwendung auf
bewegte Körper zu Asymmetrien führt, welche den Phänomenen
nicht anzuhaften scheinen, ist bekannt. Man denke z. B. an
die elektrodynamische Wechselwirkung zwischen einem Mag-
eten und einem Leiter. Das beobachtbare Phänomen hängt
hier nur ab von der Relativbewegung von Leiter und Magnet,
während nach der üblichen Auffassung die beiden Fälle, daß
der eine oder der andere dieser Körper der bewegte sei, streng
voneinander zu trennen sind. Bewegt sich nämlich der Magnet
und ruht der Leiter, so entsteht in der Umgebung des Magneten
ein elektrisches Feld von gewissem Energiewerte, welches a
den Orten, wo sich Teile des Leiters befinden, einen Str
erzeugt. Ruht aber der Magnet und bewegt sich der Lei
so entsteht in der Umgebung des Magneten kein elektris
Feld, dagegen im Leiter eine elektromotorische Kraft, w
an sich keine Energie entspricht, die aber — Gleich
Relativbewegung bei den beiden ins Auge gefaßten
vorausgesetzt — zu elektrischen Strömen von dersel
und demselben Verlaufe Veranlassung gibt, wie im
die elektrischen Kräfte.

Published in *Annalen der Physik* (1905) 17:891.

There is something that might strike you a little odd about this paper. Roughly translated, the title is "On the electrodynamics of moving bodies." Is that strange or what? It says nothing about *space* or *time* or *nuclear power* or blowing people's minds with strange ideas! Why was it that Einstein focused on electrodynamics? Obviously, he thought this was the most important aspect of the paper.

To understand what it was that most bothered Einstein as he developed special relativity, we need to think a little bit about electricity and magnetism.
TIME OUT!
Herr Professor, this book is entitled *Relativity and Quantum Physics For Beginners*. It doesn't say a thing about electricity and magnetism. Stick to the script, please!
Sigh . . . it is so difficult for me to work with the little people who don't understand my genius. We simply *must* talk about electricity and magnetism at this point.
Yeah, quit picking on my friend you brute! Where did you get that outfit anyway? It is positively horrid. You know, vertical stripes make you look fat. You should wear something slimming.

Professor Einstein, help me out here. Tell us a bit about what bothered you at the time you invented special relativity.

Man denke z. B. an die elektrodynamische Wechselwirkung zwischen einem Magneten und einem Leiter.

In English, please!

Ah, yes. Okay. Sorry. I've been dead for many years so you can't expect me to remember silly things like the language of this book. Let me see. To understand what bothered me at the time, we have to consider the electric force and the magnetic force . . .

The electric force is familiar to most of us through static cling. Often, when we rub dissimilar materials together they will stick together . . . think of rubbing a balloon on your head on a cool, dry day.

The magnetic force is familiar to most of us because we played with bar magnets. Bar magnets repel and attract each other through the magnetic force.

For the moment, let's consider just the **electric force**. According to physicists, there is an electric force between things that have electric charge. Electric charge comes in two varieties—positive and negative. The force between similar charges is repulsive, while the force between dissimilar charges is attractive.

bar magnets

Imagine a particle with positive charge sitting in space. Another particle with positive charge near the first will experience a repulsive force.

How do we best think about this? What happens in the space around the charge on the left so that the charge on the right feels a force of repulsion? I need to meditate about this.

Often in science it's useful to make up a model for how things work. It may or may not turn out to be true, but it can help us learn more about the phenomenon we are studying.

Consider a positive electric charge sitting out in space. Physicists imagine that this charge creates an electric field in the space around it. The idea of the electric field is that it is a condition such that if an electric charge is in a region of space with an electric field, that charge experiences a force!

The idea of an electric field is not a strange as it may seem to you at first. Weather forecasts are usually accompanied by a temperature map as you might see on the top to the right. This tells the temperature at different points in space. In this case, it gives the temperatures around New York. You could imagine a similar map—like that on the bottom to the right—that shows the direction and speed of the wind at different points in New York. Physicists like new words. So, instead of calling these temperature and wind *maps*, they might call them representations of temperature and wind *fields*.

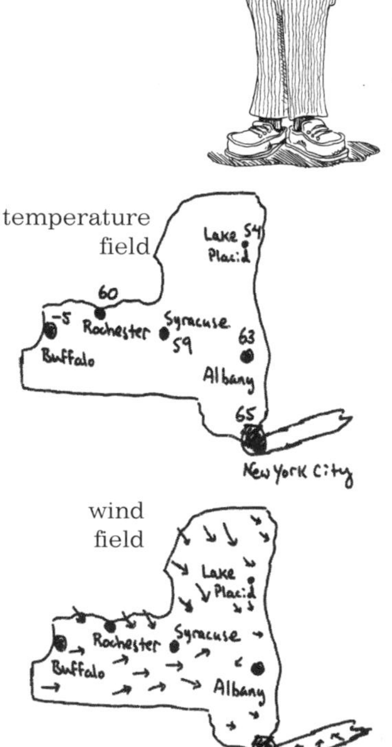
temperature field
Lake Placid 54
60
-5
Rochester
Syracuse 59
63
Buffalo
Albany
65
New York City
wind field
Lake Placid
Rochester
Syracuse
Buffalo
Albany
New York City

Physicists imagine that the presence of an electrical charge creates an electric field in the space around it. To the right is a sketch of an electric field map surrounding an electric charge. Each arrow represents the direction of the force the charge would exert on a different positive charge if it were placed in the position of the arrow. The electric field around a charge is something like a force field that repels like charges attracts unlike charges.

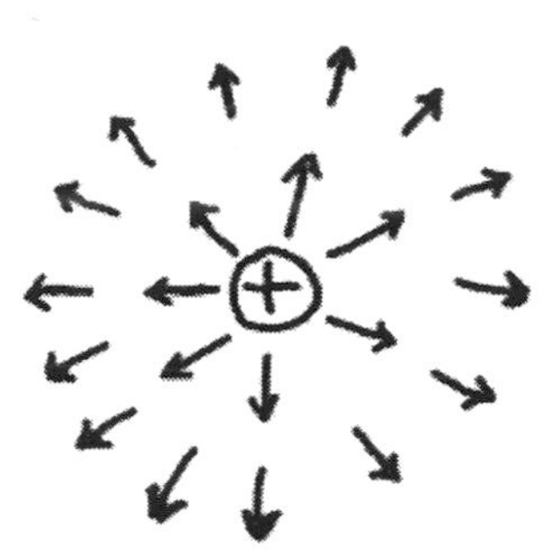

This is all fine and nice . . . but, WHY DO I CARE about this??

Be patient. Herr Professor was dropped on his head when he was young. He'll get around to the point soon.

Physicists find the idea of fields to be a very useful way to visualize and calculate the consequences of forces. So they talk about electric fields, as we've seen, and they also talk about magnetic fields and gravitational fields. Magnetic fields are the condition in space around magnets that conveys the magnetic force to other magnets. Gravitational fields are the condition in space that conveys the gravitational force to other masses.

Wait a second. I thought the force of gravity is conveyed by a warping of space-time, whatever that is.

Both gravitational fields and the warping of space-time are ways to describe the force of gravity. We have this kind of thing all the time in language. The color of blood is red. It is also the color of a ripe tomato. How you choose to describe blood depends on what you are trying to accomplish with the description.

An electric field surrounding an electric charge causes a force on another electric charge nearby. A magnetic field surrounding a little bar magnet causes a force on another bar magnet (or compass needle) nearby.

N S
bar magnet

moving charge

But things are a little more complex than this. There is a deep relationship between electricity and magnetism. A bar magnet does not move near a stationary electric charge. That means a stationary electric charge does not create a magnetic field in the space around it. However, if the electric charge is moving it *does* exert a force on a bar magnet nearby. That means a moving electric charge creates a magnetic field in space.

Okay. Now we are ready to see what it was that bothered me about electricity and magnetism.

You see, below on the left we have a charge sitting still. Under such a condition, there is an electric field but no magnetic field. If we hold a charge stationary near a compass magnet, nothing happens. But, suppose we hold the compass and go running past the charge from right to left. To us the charge would appear to be moving from left to right. This is what we see in the sketch on the right. Because the charge is moving according to our point of view, we see that the charge creates a magnetic field and that it affects a compass needle as it passes by.

N
W
E
S

positive electric charge

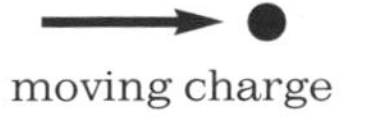

moving charge

N
W
E
S

In the one case we attribute the force to an electric field and in the other case we call it a magnetic field. Yet all that changes is our point of view—whether we are moving with respect to the charge. This means that electric fields and magnetic fields are not absolute. What you see depends on your motion! This was the problem that I solved that prompted me to entitle my paper "On the Electrodynamics of Moving Bodies." But, don't get the wrong idea. I'm not the first person to see a deep relationship between electricity and magnetism. I simply managed to understand that relationship more clearly.
Al, I'm still not feeling good about this. Have you said something that makes me care yet? Did I not hear it? Electricity and magnetism aren't doing it for me.
You should not speak to Professor Einstein that way.
Hey! Why not? He's dead.
True. Anyway, one of the reasons that electricity and magnetism are important is that they hold the key to understanding light.
Oh yeah? Please enlighten us.
Giggle, giggle. Get it? En-LIGHT-en us. Ha ha. Snort.

A hydrogen atom ran into a police station screaming, "Someone stole my electron!" A policeman asked, "Are you sure?" "Of course!" came the answer. "I'm positive."

Many physicists—including one with his picture on the $100 bill—contributed to our understanding of electricity and magnetism.

FEDERAL RESERVE NOTE
THE UNITED STATES OF AMERICA
FF 95594731 A
F6
100
ONE HUNDRED DOLLARS

*this image is not legal tender

Benjamin Franklin (1706–1790)

Michael Faraday (1791–1867)

André–Marie Ampère (1775–1836)

Joseph Henry (1797–1878)

Nikola Tesla (1856–1943)

Outside of Einstein's work explaining how to relate observations made in different points of view, physicists understood a great deal about electricity and magnetism in the latter part of the 1800s. That knowledge was pulled together in a famous series of four equations by James Clerk Maxwell in 1864. Because physicists are so original, these equations are known as **Maxwell's equations!**

Guglielmo Marconi (1874–1937)

James Clerk Maxwell (1831–1879)

My uptight friend here won't let me SHOW you Maxwell's equations. Sigh.

Maxwell's equations show that electricity and magnetism are very closely related.

In fact, physicists don't think of the electric force and the magnetic force as separate forces. Instead, they think of electricity and magnetism as different faces of a force called the *electromagnetic force*. The theoretical unification of electricity and magnetism was done by Maxwell and refined by Einstein.

I'm still not very excited.
Hey man, I'm excited! I had no idea Ben Franklin was a physicist.
Don't sweat it Surfer Dude. Nobody expects you to know much beyond the right kind of wax for your board.
What we learn from Maxwell's equations should excite you.

Maxwell's equations show us that a changing electric field (imagine a moving electric charge, for example) will create a changing magnetic field. Also, a changing magnetic field will create a changing electric field. What does this mean? If you wiggle an electric charge, it creates a changing electric field in the space around it, which induces a changing magnetic field in the space around that, which creates a changing electric field which creates a changing magnetic field, and so forth. This phenomenon of fields inducing fields moves outward at the speed of light. This is the fundamental essence of light!

Okay. Now I'm excited.

Physicists use Maxwell's equations to show that, mathematically, the electric and magnetic fields in light are *waves*.

What's that supposed to mean?

> Like as the waves make towards the pebbl'd shore
> So do our minutes hasten to their end.
> — William Shakespeare, Sonnet LX

Here's a wave for you.
Uh. Not that kind of wave.
What is a wave?
ATOMIC POP
SCIENCE WEEKLY
QUANTUM AIRLINES

Nature is full of waves. For example, there are water waves, sound waves, and waves moving on strings.

Imagine being at a big football game. The stadium is packed. Suddenly the crowd begins to do the wave. What exactly is it that makes the wave in the crowd?

Each person in the wave stays at their seat. What makes the wave happen is that the people in a given section stand in unison just after people to one side and just before people to the other side.

All waves in nature are similar to the stadium wave. For example, imagine you and a friend are holding a jump-rope so that is has a little slack.

Now you move your hand holding the rope straight up and down repeatedly. You'll see waves traveling down the rope. Each bit of rope moves straight up and down. Yet the wave pattern moves along the rope due to the timing of the movement of each part of the rope.

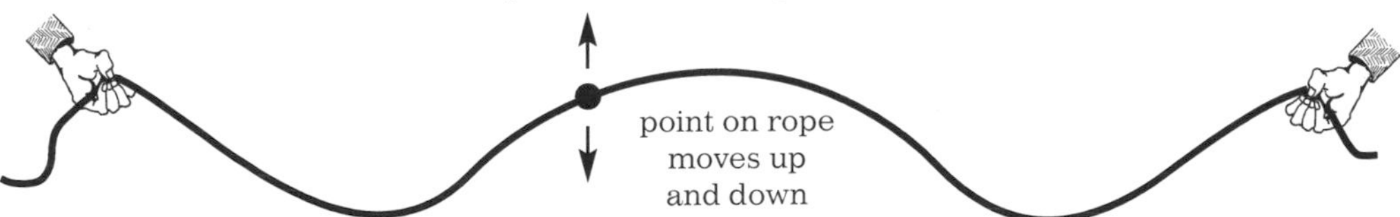
point on rope
moves up
and down

All waves in nature are similar to stadium waves and waves on a rope. What differs is what does the waving. In stadiums, people do the waving. For waves on a rope or a string—say on a guitar or a violin—the string does the waving. For sound waves, molecules in the air do the waving. In the case of water waves on a pond, the water does the waving.

All waves have a very similar mathematical description in physics. They share many of the same characteristics.

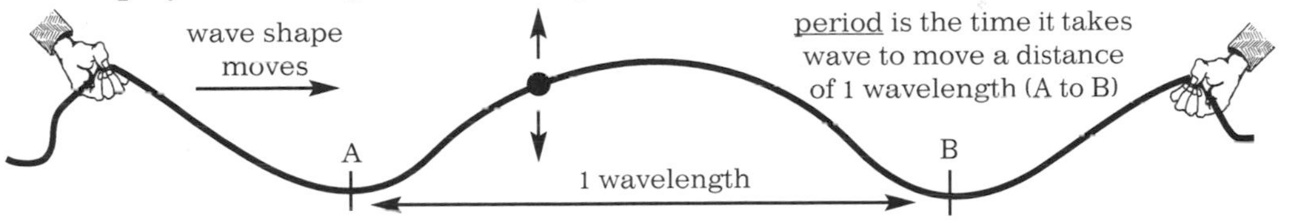
wave shape
moves
period is the time it takes
wave to move a distance
of 1 wavelength (A to B)
A
B
1 wavelength

Sound waves with different wavelengths (frequencies) are perceived differently by humans. We hear waves with shorter wavelengths as having a higher pitch.
Dude! I prefer water waves. I don't think I can surf on a jump rope.

Maxwell discovered that light is made up of electric and magnetic fields that are described by a wave equation. Light is a wave where the electric and magnetic fields do the waving. We can't see the electric and magnetic fields directly any more than we can see the air molecules move in a sound wave. Yet we perceive the air waving as *sound* and the electric (magnetic) fields waving as *light*.

In the same way that there are sounds with pitches that are too high or too low for humans to hear, there is light that has wavelengths our eyes do not perceive. Light that the human eye perceives is called *visible light*. Different wavelengths of visible light are what humans see as different colors. The wavelength of visible light is roughly 150 times smaller than the width of a human hair.

Light is usually a word we use to refer to visible light. But, we know there are similar waves with very different wavelengths. Radio waves, gamma rays, X-rays, microwaves, and infrared light are all the same type of beast as visible light except for the fact that they have different wavelengths. For example, radio waves have wavelengths ranging from a few centimeters to many thousands of kilometers. X-rays have wavelengths somewhere between a thousandth to a millionth the width of a human hair.

Physicists use the terms *electromagnetic waves* or *electromagnetic radiation* to refer generally to all these different types of "light" waves. These terms have the advantage of sounding much cooler than just "light."

The variety of electromagnetic waves

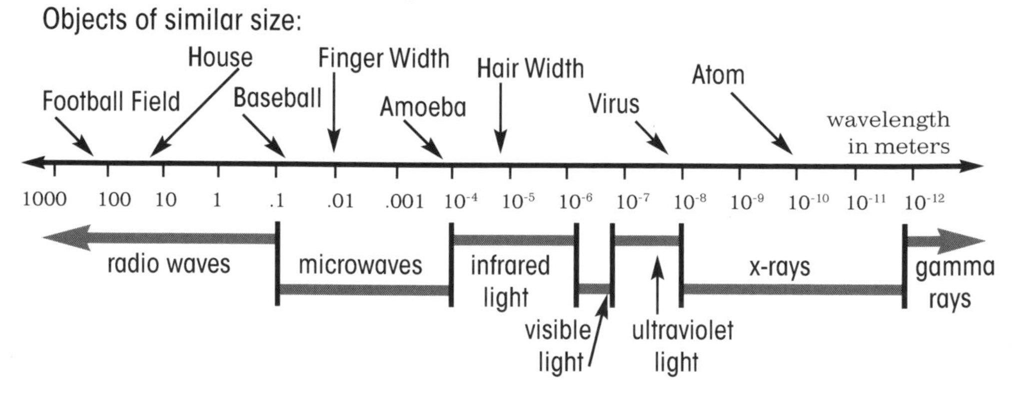
Objects of similar size:
Football Field
House
Baseball
Finger Width
Amoeba
Hair Width
Virus
Atom
wavelength in meters
1000 100 10 1 .1 .01 .001 10^{-4} 10^{-5} 10^{-6} 10^{-7} 10^{-8} 10^{-9} 10^{-10} 10^{-11} 10^{-12}
radio waves
microwaves
infrared light
visible light
ultraviolet light
x-rays
gamma rays

Chapter 8
Quantum Mechanics Light

Wow! I thought this book was going to be lame. But I learned something very cool.
Of course my child. I have taught you that pudding is the meaning of life.
No! Not *that!* You know? You are a very strange, scary, hairy old man. I learned that light is a wave! By the way, do you ever manage to take a shower? How do you do that up on top of the mountain? What do you eat?
Actually, sport. I hate to break it to you. But that whole "light is a wave" thing you are excited about? It's really not as simple as all that.
After all this, you're saying light is NOT a wave? Sheesh!
Light is definitely a wave phenomenon, but ...

In science, nature always has the final say. The wave model of light is extraordinarily successful. It underlies our understanding of optics and radios and cell phones and rainbows and optical fibers and microwave ovens and . . . But . . .

BUT what?!

But, as successful as the wave model of light is in explaining all sorts of things, it fails miserably when explaining a few other things. For example, physicists have never been able to use the wave theory of light to understand how light is absorbed and emitted by matter.

Hmmm. The fireplace poker is glowing and the glow becomes bluer as the poker becomes hotter and more red as it cools down.

Max Planck

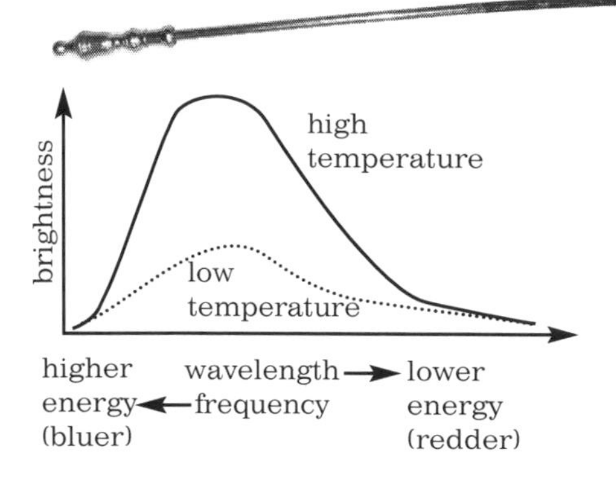
brightness
high temperature
low temperature
higher energy (bluer)
wavelength
frequency
lower energy (redder)

At the end of the 19th century, a seemingly simple aspect of the interaction of light with matter that physicists needed to explain was the color of light emitted by a glowing object. Imagine, for example, a fireplace poker glowing after being heated in the fire. To isolate the light emitted by an object—as opposed to light reflected from the object—it was necessary to put the object in a dark container (to avoid the reflected light) and record the color and brightness of the emitted light as the temperature of the object was varied. Because the object under study was kept in the dark, the light emitted by the object, the "glow" if you wish, was called *black-body radiation.*

In 1900, a German physicist named Max Planck created a theory that perfectly described the color and brightness of light emitted by objects. Strangely, though, his theory did not view light as a wave, but rather assumed that light consists of little packets (or particles) of energy.

Though Planck managed to discover a formula that described very well the color of light emitted from materials, few people took seriously his idea that light existed in the form of little packets of energy.

I have uncovered a great surprise! Light acts like a particle.

Dude! Don't get all that excited. You don't really know what you're doing. You just stumbled onto the correct final equation by accident. EVERYBODY knows light is a wave, not made up of these silly packets.

Did you know that Max Planck was awarded the 1918 Nobel Prize in Physics for "the services he rendered to the advancement of physics by his discovery of energy quanta"? Interestingly, Planck had been advised by one of his physics professors to avoid physics as a career on the grounds that everything had been discovered already!

Another development that played a role in modifying our concept of light occurred in Germany in the late 1880s when a physicist named Heinrich Hertz observed that light incident on a metal surface ejected electrons from the metal. He studied how the flow of electrons from the metal varied with the color and the brightness—or intensity—of the light.

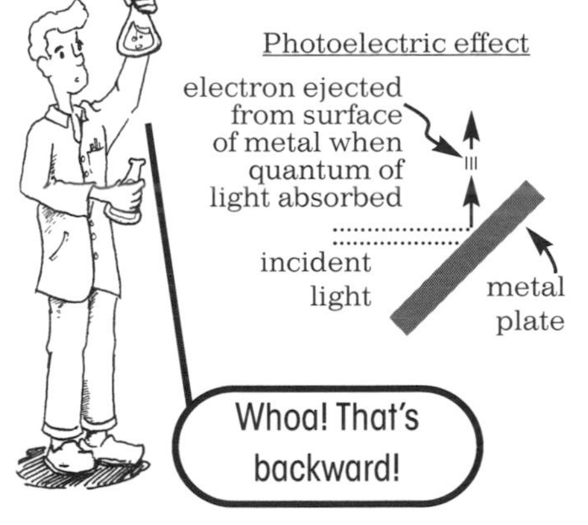
Photoelectric effect
electron ejected from surface of metal when quantum of light absorbed
incident light
metal plate
Whoa! That's backward!

According to the wave theory of light, physicists expected the energy of the electrons leaving the metal to vary with the brightness of the light, but not the color. Yet they found that the energy of the emitted electrons varied with color, *not* the brightness, of the light.

In 1905, Einstein published a paper that explained the results of these so-called photoelectric effect experiments. In his paper, Einstein was able to explain the results seen in the experiments very nicely. In doing so, however, he assumed light existed in little bundles of energy, in seeming contradiction to the wave picture. In 1915, the physicist Robert Millikan did careful photoelectric effect experiments and showed that the details of Einstein's theory were correct.

You're kidding! Not that Einstein guy again. Incredible. He must've needed a hobby.

Millikan's work also demonstrated that the packets of light energy seen in the photoelectric effect were *exactly* like the ones needed to explain black-body radiation.

Two completely different types of phenomenon—black-body radiation and the photoelectric effect—were explainable only if light was assumed to come in little packets with an energy that varied with color. In particular, the energy of each packet was found to be equal to a constant multiplied by the frequency of the light. (Recall that frequency is a way of measuring the wavelength, or color, of the light.) Experimenters determined the constant in both cases and found it to be the same! It is called *Planck's constant* and physicists symbolize it by h.

With two independent phenomena and sets of experiments pointing to the same conclusion, the physicists of the world were unable to avoid the conclusion that light consists of little packets—or *quanta* (*quantum* for an individual packet)—of energy.

Well *that* makes sense! h stands for Planck's constant. Didn't anyone think to use *p* to stand for *Planck*? What's wrong with you guys? And I'm the one with the rep for being dumb. Go figure.
It's sort of like discovering this awesome dress for the prom and then seeing someone else wearing the exact same dress!

Maybe. I can't say. Nobody had a dress like mine at the prom! Ha, ha. Just kidding.

An imperfect analogy might help to put Einstein's work on the photoelectric effect in perspective. Back in September 2001, when the first plane struck the North Tower of the World Trade Center, many people naturally thought it was a horrible accident. When, a short while later, the South Tower was struck by a plane, those same people knew with absolute certainty that the awful disaster was no accident. Though the scientific shock in 1905 was not a tragic one, the shift in thinking was abrupt in the same sense. Think of Planck's work on black-body radiation as the first tower being struck, and Einstein's work as the second tower being struck. Suddenly scientists *knew* the particulate nature of light was here to stay. Happily, the analogy ends there.

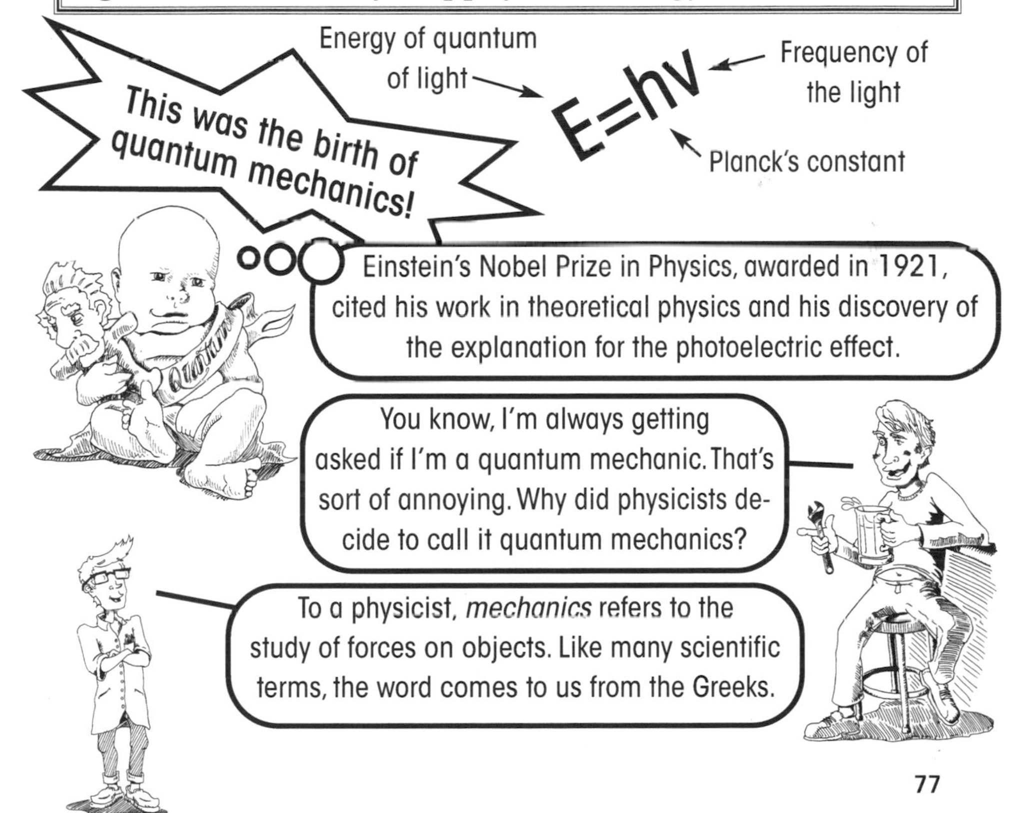
This was the birth of quantum mechanics!
Energy of quantum of light
E=hv
Frequency of the light
Planck's constant
Quantum
Einstein's Nobel Prize in Physics, awarded in 1921, cited his work in theoretical physics and his discovery of the explanation for the photoelectric effect.
You know, I'm always getting asked if I'm a quantum mechanic. That's sort of annoying. Why did physicists decide to call it quantum mechanics?
To a physicist, *mechanics* refers to the study of forces on objects. Like many scientific terms, the word comes to us from the Greeks.

Heinrich Hertz
(1857–1894)

Remember that guy Hertz who discovered the photoelectric effect? In spite of the importance of the photoelectric effect in the eventual development of quantum mechanics, Hertz was best known for his work with electromagnetic waves. He established the existence of radio waves that behaved exactly as Maxwell predicted they should. The unit of frequency for electromagnetic waves is called the *hertz* in honor of this man. If you look on an old radio dial (and some new ones), you'll find the symbols kHz and MHz. These stand for kilohertz (thousand hertz) and megahertz (million hertz), respectively. Sadly, Hertz died at age 36. Oddly enough, his research contributed toward the establishment of both the wave picture and the particle picture of light.

So, now you're telling me light is made of particles (or quanta). Obviously, if this is the case, light cannot be a wave. So Maxwell was wrong and this book is full of lies, lies, lies!

> It's of no use whatsoever.
>
> —Heinrich Hertz, when asked about the usefulness of his discovery of radio waves.

Physicists were rather bothered by the apparent contradiction. Smooth, continuous waves and particles seem like very different beasts. How can light have characteristics of both waves and particles?

Light is a wave!
Light is a particle!
Excuse me. I know I shouldn't interrupt since I'm just a stupid looking cartoon character and not an eminent dead scientist, but perhaps you are ALL correct. Light is both a wave and a particle.

Why is it that humans are bothered about this wave-particle thing? I can't see the particles or the waves by just looking at this beam of light. So what difference does it make?

Dude! I ride waves all the time, and I never saw one that was particles. You can't have both at the same time.

Things like water waves and waves on strings have wavelike characteristics but don't act at all like particles. Marbles flying through the air act like particles, not like waves. Physicists have good theoretical models of waves and of particles. Both waves and particles exhibit very particular properties. The world of the human experience sees these two ideas as distinct conceptually. We do not have the ability to easily imagine something that has both wave and particle characteristics.

Humans are uncomfortable when a black-and-white picture gets muddled. Murder is bad. Saving lives is good. Would it have been right to murder Hitler in 1938, possibly avoiding World War II and saving many lives in the process? Some of us would say yes and others no. Most of us feel uncomfortable thinking about it.

Nature has no hang-ups about particles versus waves. Light is what it is. We find that light has both particle characteristics and wave characteristics. That is what we've learned through the scientific process. Light appears to have what is often called *wave-particle duality*, which is to say it can act like a wave or a particle.

> And these fifty years of conscious brooding have brought me no nearer to the question of "What are light quanta?" Nowadays every clod thinks he knows it, but he is mistaken.
>
> —Albert Einstein, shortly before his death

As with relativity, what we've discovered about light demands that humans look beyond the familiar comfort zone. Depending on what question you ask, light is a particle-like thing or a wave-like thing. The scientific model used for light depends on whether the particle characteristics or the wave characteristics are important for answering the question at hand. There is no reason to expect nature to conform to our comfort zone.

> Every great advance in science has issued from a new audacity of imagination.
> —John Dewey

A famous chemist named Gilbert Lewis proposed a theory about quanta of light in which he called them *photons* (a word similar to the Greek word for light). Gilbert's theory didn't last long, but the word *photon* stuck. Photons are what physicists commonly call the quanta of light initially proposed to exist by Einstein and Planck.

This is VERY important, of course. Why? Well, because the word *photon* sounds really cool!

> Light and matter are both single entities, and the apparent duality arises in the limitation of our language. —Werner Heisenberg

Isaac Newton (1634–1727)

Way back in 1670, Isaac Newton tried to explain some of light's characteristics by assuming it was made of little particles he called *corpuscles*. Maxwell's electromagnetic waves explained these things well enough though, and the light-as-particle idea fell out of favor.

Chapter 9

What's the Matter?

Okay. This whole wave-particle thing is making my head explode.
Yeah. It's sorta like the first time my sister came home with a date and I had to think of her both as my sister AND as a girl. I mean THAT was weird!
Er. Uh. Right, Surfer Dude. Whatever you say.
I've got some more bad news for you guys, It gets worse. There once was a guy named Niels Bohr and another by the name of de Broglie ...
Actually, the latter's name was Louis Victor Pierre Raymond duc de Broglie. Imagine being saddled with THAT name!
ΦBK
No, no, no! We're not ready for that yet you nincompoop! First, we must set the stage! We have to make the audience feel the part, yes? Tell them about the atom BEFORE Bohr and de Broglie!

At the start of the 1900s, physicists knew that atoms were made of little particles with a positive electric charge called *protons* and even smaller negatively charged particles called *electrons*. But they didn't know how these particles fit together in the atom.

Plum Pudding

Nuclear

How do you see inside an atom? I mean it's too small to see, right?

Imagine that you didn't know how a car worked and you weren't allowed to open the hood to see inside. How might you learn something about the car?

I'd just Google "car" and look at the pictures and the explanation.

You could take a rifle and shoot numerous bullets at the car. Bullets hitting the passenger compartment would pass through the car easily, because there is very little material to stop them. Bullets hitting the engine area, however, would be unable to pass through—they would ricochet off the engine. By looking at the way in which bullets scatter from the car as you shot it many times, you could infer the fact that most of the material in the car is located in the engine area.

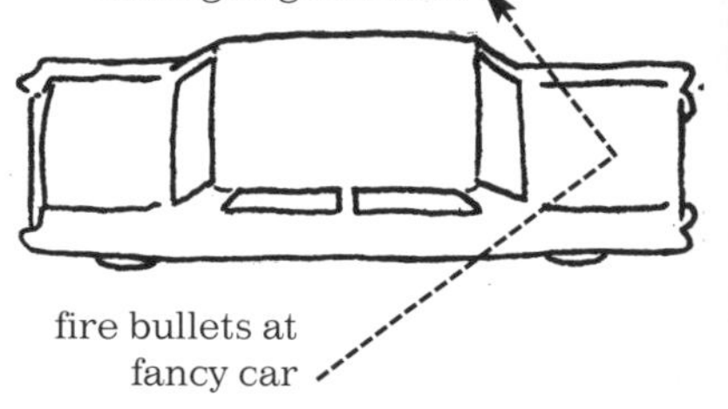
bullets deflected when hitting engine block
fire bullets at fancy car

This may seem sort of silly, but it is very similar to what physicists did to first "see" inside the atom.

In 1900, the leading model of the atom was known as the *plum-pudding model*.

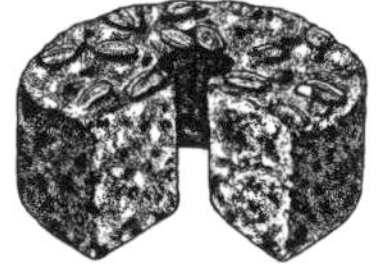

In the plum-pudding model of the atom, the negative electrons are thought to be embedded in a cloud of positive charge in the same way that bits of fruit might be suspended if they were tossed into the bowl and mixed into the pudding.

I keep telling you that pudding is the meaning of life.

In 1909, Hans Geiger and Ernest Marsden, working under Ernest Rutherford in a lab at the University of Manchester in England, shot "bullets" at gold atoms and observed that much of the time the bullets passed right through the gold foil, while once in a while the bullets bounced backward, much like the bullets ricocheting off the engine block of the car.

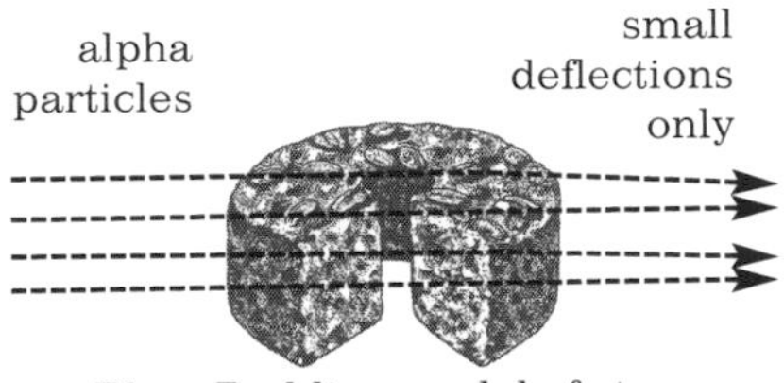
alpha particles
small deflections only
Plum Pudding model of atom

The bullets fired by Geiger and Marsden at the gold atoms were positively charged particles called *alpha particles*, which are emitted by certain types of materials.

If you *really* want to know, alpha particles are made up of two protons and two neutrons and are naturally emitted by some types of unstable atoms. It's a form of natural radioactivity.

When an alpha "bullet" passes through a gold atom, the positive alpha is repelled or deflected by the positive charge in the atom. If the positive charge in the gold atom is spread out—as expected in the plum-pudding model of the atom—the mass and charge of the atom can only deflect the alpha particles a small amount. It is sort of like shooting a car with no engine—the bullets always go through the car.

Ernest Rutherford (1871–1937)

Rutherford and his colleagues expected to see small deflections of the alpha bullets, thus proving the plum-pudding picture. They did see many alphas deflected by small amounts, but were surprised to find some alphas bouncing backward from the foil.

Why, yes! Rutherford's friend Hans Geiger IS the same guy who invented the Geiger counter, an instrument that is widely used to detect radioactivity.

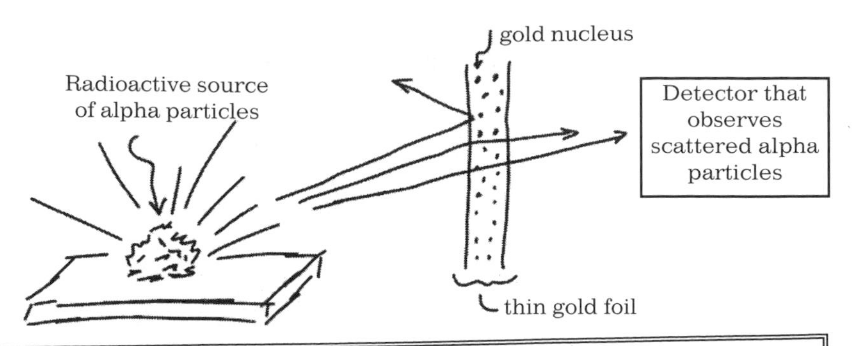
gold nucleus
Radioactive source of alpha particles
Detector that observes scattered alpha particles
thin gold foil

It was quite the most incredible event that has ever happened to me in my life. It was almost as incredible as if you fired a 15-inch shell at a piece of tissue paper and it came back and hit you. —Ernest Rutherford

Rutherford realized that if most of the mass and all of the positive electric charge in the atom were concentrated in a small ball at the atom's center, most of the time the alpha bullets would pass through the atom, and once in a while the alpha bullet would be deflected backward.

So Rutherford suggested that the positive protons in the atom were concentrated in a nucleus at the center of the atom. This is what physicists call the nuclear model of the atom.

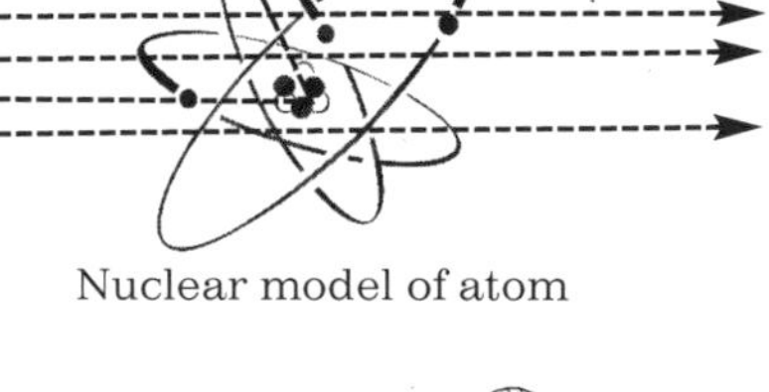
alphas
Nuclear model of atom

A young Danish physicist named Niels Bohr worked in Rutherford's laboratory during the time that Rutherford, Geiger, and Marsden did their scattering experiments. He understood the significance of the work being done and proposed a famous nuclear model of the atom in 1913. This model was important in the development of quantum mechanics, because it was the first instance in which the structure of the atom was quantized.

Niels Bohr (1885–1962)

Ernest Rutherford won the Nobel Prize in Chemistry in 1908 for his "investigations into the disintegration of the elements, and the chemistry of radioactive substances."

In the Bohr model of the atom, the protons and neutrons are concentrated in a tiny little nucleus at the atom's center while the electrons move in circular orbits about the nucleus.

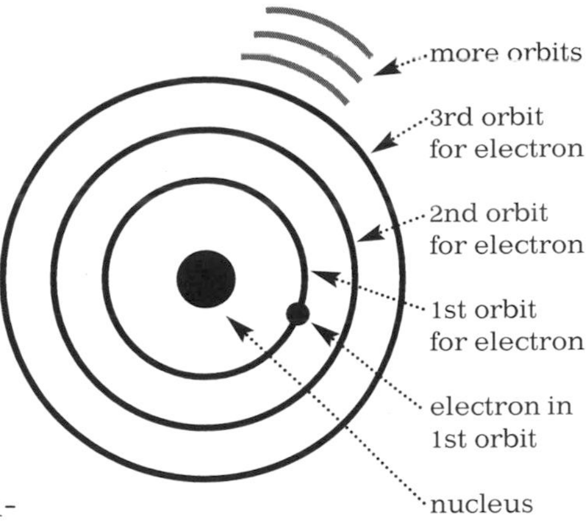

more orbits
3rd orbit for electron
2nd orbit for electron
1st orbit for electron
electron in 1st orbit
nucleus

Since the protons and the neutrons are massive compared to the electrons, most of the mass in the atom is concentrated at the center. The positive electric charge is also concentrated in the nucleus, because that's where all the protons reside.

Man is this bohring?! Get it? BOHR-ing. Ha ha!

The positive nucleus attracts the negative electrons and holds them in the atom.

The electron in the atom is a bit like a rock being twirled by a person using a sling. The leather of the sling holds the rock near the person until he releases the sling. In the Bohr model of the atom, the electrical attraction of the negative electrons to the positive nucleus acts like the leather, holding the electron in the atom.

Bohr proposed something very new, very important, and very strange in his atomic model: he "quantized" the electron orbits!

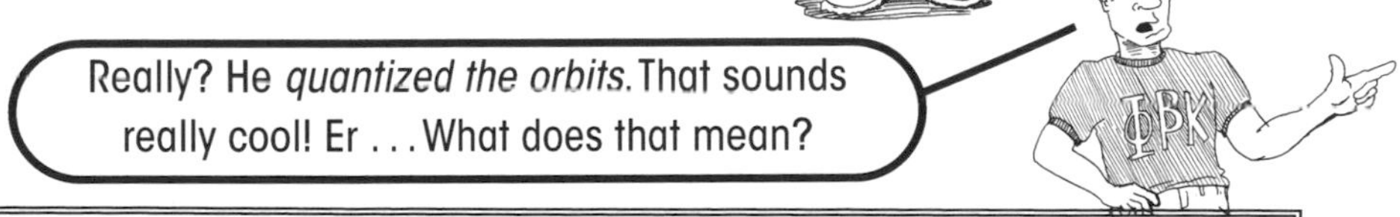

Really? He *quantized the orbits.* That sounds really cool! Er . . . What does that mean?

> Prediction is very difficult, especially if it's about the future. —Niels Bohr

In the normal, so-called classical, physics of the day, an electron moving in a circle emits light and loses energy. As it loses energy, it spirals into the nucleus, and the atom is no more. The light that's emitted comes out in a wide band of colors. This doesn't happen with real atoms.

You see, the classical picture simply didn't work! Atoms are stable, and the positive charge is all concentrated in the center. Also, only very particular colors of light are absorbed and emitted by atoms. I had to think of something new.

Bohr proposed that the electron could exist only in very particular orbits in the atom—meaning the orbits were "quantized"—and that these orbits were stable for some unknown reason. He then hypothesized that the electron could make transitions between different stable orbits by emitting or absorbing a packet of light, a *photon*.

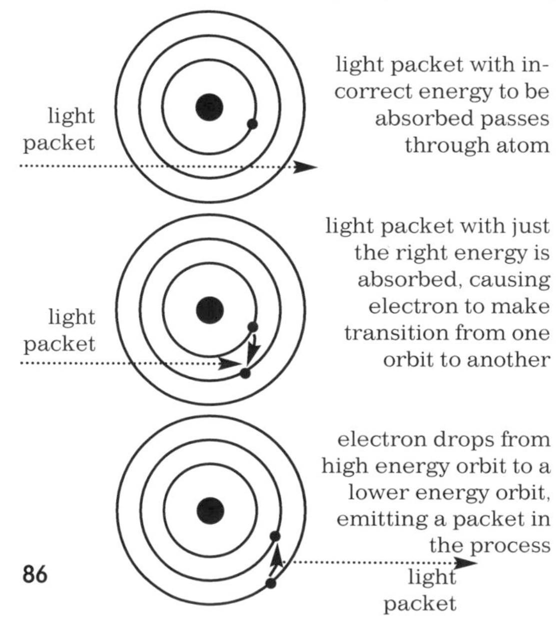
light packet
light packet with incorrect energy to be absorbed passes through atom
light packet
light packet with just the right energy is absorbed, causing electron to make transition from one orbit to another
electron drops from high energy orbit to a lower energy orbit, emitting a packet in the process
light packet

Niels Bohr won the Nobel Prize in Physics in 1922 "for his services in the investigation of the structure of atoms and of the radiation emanating from them." Niels' son, Aage Bohr, also won the Nobel Prize in Physics. Imagine the conversations over cereal in THAT house!

Bohr's model of the atom was the first atomic model that successfully described *atomic spectra*, which is the pattern of the color of light emitted and absorbed by matter. It had been known for many years that each type of atom emits and absorbs light at only a discrete set of colors (or frequencies).

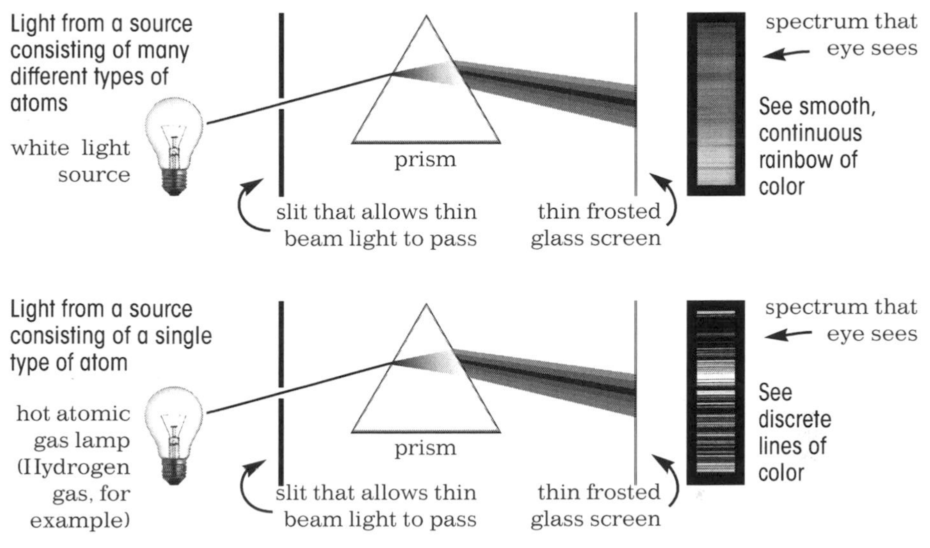

Light from a source consisting of many different types of atoms
white light source
prism
slit that allows thin beam light to pass
thin frosted glass screen
spectrum that eye sees
See smooth, continuous rainbow of color
Light from a source consisting of a single type of atom
hot atomic gas lamp (Hydrogen gas, for example)
prism
slit that allows thin beam light to pass
thin frosted glass screen
spectrum that eye sees
See discrete lines of color

Different atoms at source give different line patterns on screen

The pattern of colors emitted by each type of atom acts like a fingerprint for that atom. Bohr's model predicted the electron could exist only in discrete (quantized) orbits and that only light with a photon energy or color that corresponded to exactly the difference in energy between two Bohr orbits could be emitted or absorbed. Bohr's model was able to describe exactly the colors emitted by simple atoms (those with one electron) by specifying the exact energies of the allowed electron orbits. This was an amazing scientific success!

I got fingerprinted once by the police. I tell you, nobody thought *that* was a scientific success—especially my wife!

> How wonderful that we have met with a paradox. Now we have some hope of making progress. —Niels Bohr

In the Bohr model, the idea of the photon is built into the model. The fact that the Bohr model was able to successfully describe atomic spectra was yet *another* big reason to believe in the particulate nature of light!

Okay. Maybe this isn't so Bohring after all!
It certainly isn't boring. Just wait until you hear what happened next!

In 1924, a French doctoral student made a stunning suggestion. Louis de Broglie proposed that if light could be both a wave and a particle, then perhaps matter particles, like electrons, could also be both waves and particles! He predicted how the "wavelength" of matter depends on the mass and velocity of the particle.

Louis de Broglie
(1892–1987)

De Broglie's idea provided an interesting explanation for why only certain electron orbits are stable in the Bohr picture of the atom. Imagine an electron with a certain "wavelength." Only orbits with a circumference that is a multiple of the wavelength for the electron would be stable, otherwise the electron wave would cancel itself out as it goes around the circle! This turns out to be very similar to the reason that vibrating strings on guitars and violins play only certain notes clearly.

start

unstable orbit for electron

wave would cancel itself out as it travels around and interferes with itself

start

stable orbit for electron

wave would reinforce itself as it travels around and interferes with itself

Wait a minute! Waves are very different from particles. If, all of a sudden, we decide that particles are waves instead of particles, all our physics theories don't work! Right?

As a matter of fact . . . yes, that's sort of true. For normal-sized objects, the wavelength is too short to notice, and we don't need to take the wave nature of the object into account. Classical physics works fine in this case. But for very small things—vastly smaller than you can see with your eye—the wave nature of the object is very important, according to de Broglie. Back in the 1920s, physicists realized this and had to invent a new type of physics to deal with matter waves.

This new type of physics is what we know as quantum mechanics.

De Broglie (pronounced *debroy*) made his hypothesis about matter waves as part of his doctoral thesis. He was awarded the 1929 Nobel Prize in Physics for "his discovery of the wave nature of electrons," making him the first person to receive the Nobel Prize for a PhD thesis.

Since 1924, there have been many scientific experiments showing that matter has wavelike properties. Wave-particle duality appears to be Nature's way in spite of our difficulty in visualizing it.

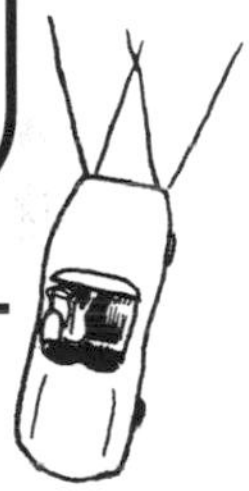

Still, unless I'm Surfer Dude, I don't need to worry about waves on my way to work, right? I mean this isn't stuff we normally have to think about.

Right. Tiny things like electrons often act like waves, but baseballs and cars just act like big particles. Baseballs and cars have wavelengths, as strange as that seems, but their wavelengths are way too small to notice.

> Thus to describe the properties of matter as well as those of light, waves and corpuscles have to be referred to at one and the same time. The electron can no longer be conceived as a single, small granule of electricity; it must be associated with a wave and this wave is no myth; its wavelength can be measured and its [properties] predicted.
>
> —Louis de Broglie in his Nobel lecture (1929)

Electrons are very small, so the wave nature of the electron is important. In order to understand the atom, physicists had to invent a theory that took into account the wave nature of the electron. Many famous physicists contributed to this effort.

Some of the major players in the early development of quantum mechanics

Max Born (1882–1970) won the 1954 Nobel Prize in Physics for "his fundamental research into quantum mechanics."

Paul Dirac (1902–1984) won the 1933 Nobel Prize in Physics for "the discovery of new productive forms of atomic theory."

Werner Heisenberg (1901–1976) won the 1932 Nobel Prize in Physics for "the creation of quantum mechanics."

Erwin Schrödinger (1887–1961) won the 1933 Nobel Prize in Physics (with Dirac) for "the discovery of new productive forms of atomic theory."

Wolfgang Pauli (1900–1958) won the 1945 Nobel Prize in Physics for "the discovery of the Exclusion Principle."

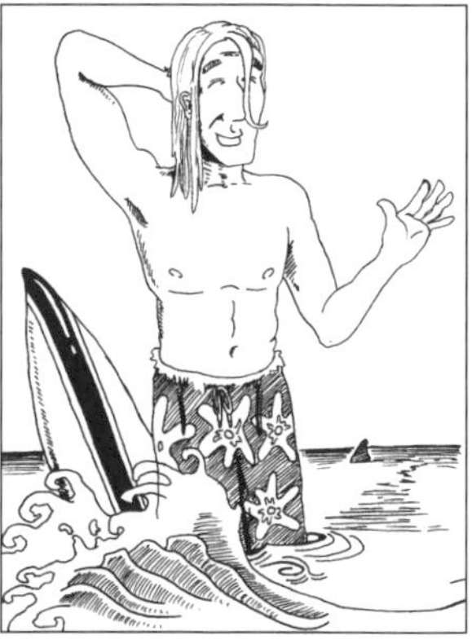

Surfer Dude (1989–) not expected to win the Nobel Prize in Physics, but did eat three servings of high school cafeteria casserole once and lived to tell the story.

Okay Surfer Dude. Let's go. You aren't allowed on this page.

During the 1920s, physicists developed, and slowly understood, the mathematical relationships that govern the behavior of small particles, taking into account the newly discovered wavelike aspect of matter.

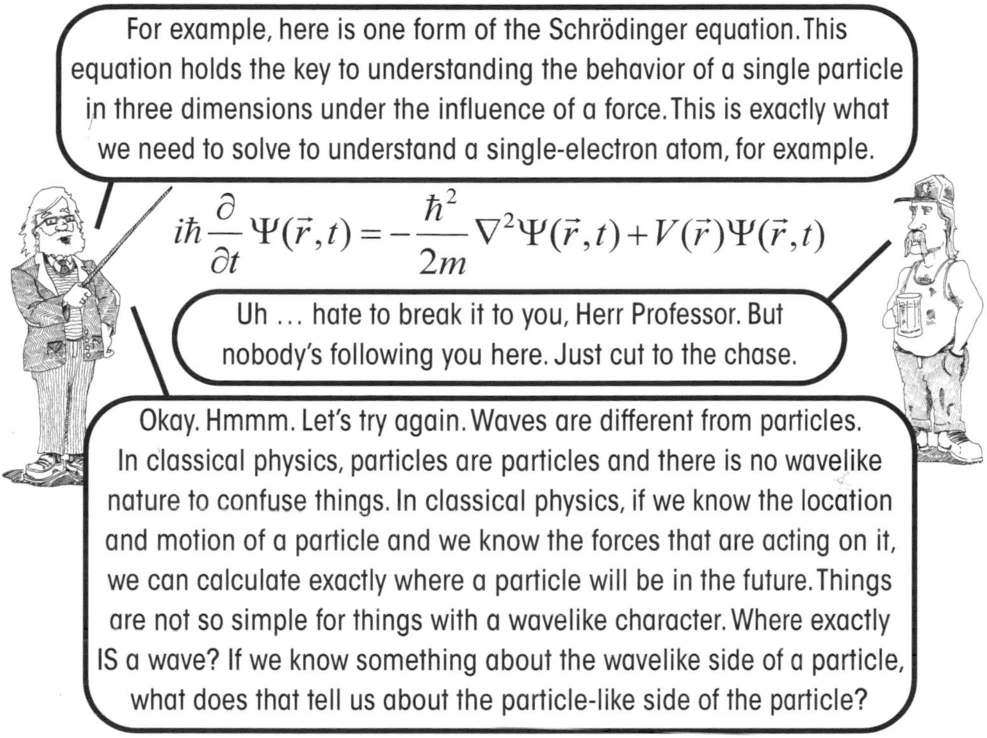

For example, here is one form of the Schrödinger equation. This equation holds the key to understanding the behavior of a single particle in three dimensions under the influence of a force. This is exactly what we need to solve to understand a single-electron atom, for example.

$$i\hbar\frac{\partial}{\partial t}\Psi(\vec{r},t) = -\frac{\hbar^2}{2m}\nabla^2\Psi(\vec{r},t) + V(\vec{r})\Psi(\vec{r},t)$$

Uh ... hate to break it to you, Herr Professor. But nobody's following you here. Just cut to the chase.

Okay. Hmmm. Let's try again. Waves are different from particles. In classical physics, particles are particles and there is no wavelike nature to confuse things. In classical physics, if we know the location and motion of a particle and we know the forces that are acting on it, we can calculate exactly where a particle will be in the future. Things are not so simple for things with a wavelike character. Where exactly IS a wave? If we know something about the wavelike side of a particle, what does that tell us about the particle-like side of the particle?

Confronted with the wavelike nature of matter, physicists found that they could not analyze the motion of particles under the influence of forces as they had done since the days of Newton. For large particles, like dogs and cars and so forth, the wavelike nature is not noticeable, and the old way of doing things works fine. But for tiny particles, where the wavelike character of matter is important, the old ways are not sufficient to understand things. Quantum mechanics was invented to solve this problem. The thing is, though, what comes out of quantum mechanics seems a little strange to most of us. What you get out of quantum mechanical calculations are probabilities rather than certainties. You can't say precisely where a particle is located, but rather you determine the probability that it is located in a certain region of space.

Take the atom as an example:

Consider the simplest atom where there is one electron attracted to a proton.

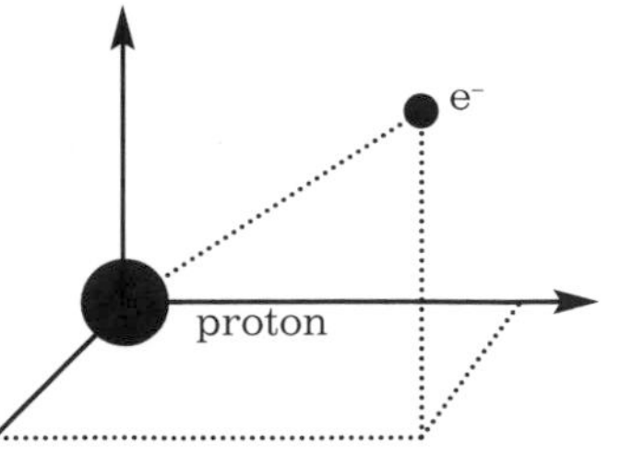
e^-
proton

Oh! Look honey! It's a hydrogen atom! Can we take it home?
Now Ethyl. We have other atoms at home. They'll be jealous. Besides you know the kids won't take care of it.

Similar to the Bohr model of the atom, the quantum mechanical solution of the single-electron atom tells us that the electron can only exist in discrete states.

Red states or blue states? Northern or Southern? Don't electrons exist in each of the 50 states?
ΦBK
Uh . . . not THAT kind of state.

In quantum mechanics, when physicists refer to *states*, they really mean *states of existence*. For example, the quantum mechanical solution of the single-electron atom says that the electron can exist only in particular spatial regions around the nucleus and with a particular set of energies. An electron in one of these spatial regions with a particular acceptable energy is said to be in a specific *quantum state*. The Bohr model is similar in that the electron only exists in circular orbits, each of which corresponds to the electron having a different, fixed energy.

Quantum states in an atom are a bit like houses lining a street that meanders up a mountain. A person can live in each house, but not between. A person with a lot of energy can hike up to the higher houses and live, while a person without much energy must make do with a lower house. In the world of quantum mechanics, each available house is analogous to each available quantum state in the atom.

Atomic
Hills Ski
Resort

Unlike the simple circles of the Bohr model, the full-blown quantum mechanical solution to the atom has strangely shaped regions where the electron can be found. Here are examples of the shapes of the available quantum states in the atom. These shaded regions represent areas where the electron would most likely be found around the nucleus if it were in that particular quantum state, or available house. Think of these as a few of the available house plans.

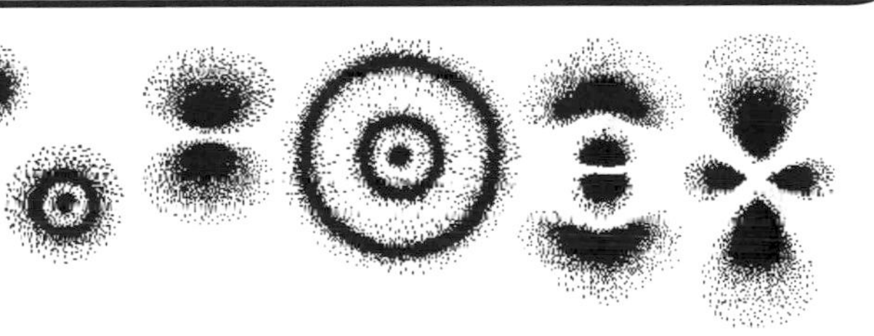

Most atoms have multiple electrons. Typically in these atoms—according to quantum mechanics—the electrons fill the available quantum states from the lowest energy state upward, with two electrons per state. The spatial distribution and the energy of the electrons is determined by the charge on the nucleus and the number of electrons in the atom.

Different types of atoms (known as *elements*) have different amounts of positive charge in the nucleus and different numbers of electrons. This means the spatial configuration and energies of the electrons around the nucleus in each type of atom is unique.

Each atom is called an element?
Oh no! Each *type* of atom is an element. So, for example, gold is an element. So are iron and carbon and oxygen. There are about 100 elements in all. Each element has a unique set of characteristics and way of interacting with other elements, which is what a scientist would call the *chemistry* of the atom.
Heh, heh, heh. Now I know you're joshing me. See, there's way more than a hundred different types of materials in the world.
Sigh. You need to go back to watching your ballgame Fred. Haven't you ever heard of a *molecule*?

Atoms interact with each other through rearrangements of the electrons that surround them.

Elements each have different chemical properties because the electrons around the nucleus differ in number and arrangement. Atoms can also interact with one another to form combinations of atoms called *molecules*. For example, carbon dioxide is a molecule in which carbon is bonded to two oxygen atoms. Molecules can have physical properties (appearance, taste, hardness, and so forth) and chemical characteristics that are different from the elements that form them. This is how we have so much variety in the world even though there are only slightly more than 100 different elements.

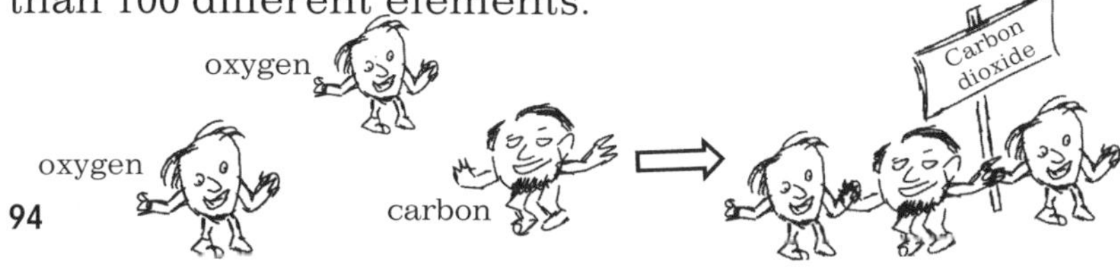
oxygen
oxygen
carbon
Carbon dioxide

But why is it some atoms join together into molecules and others don't? I really wouldn't want my gold bracelet to turn into some funky gold dioxide gas and waft away!

Many things determine when chemical reactions occur and when they don't. Chemists are the experts in that. Typically, atoms will bond or trade partners when it is energetically favorable to do so. That means it happens if the end products have less energy tied up in them than the initial materials.

Sometimes it takes a little kick to get things started. For example, wood burns, but it always needs just a little help to get started. The heat and light produced by the burning wood is the energy released as the molecules in the wood are reconfigured to smaller molecules that hold less energy.

Let's look at an example.
Sodium Ln.
Chlorine St

Sodium is a grayish metal that is explosive. An atom of sodium has 11 electrons. According to quantum mechanics, they would get arranged in our mountain village as shown in the sketch.

Chlorine is yellow gas that is corrosive. An atom of chlorine has 17 electrons that are arranged as shown.

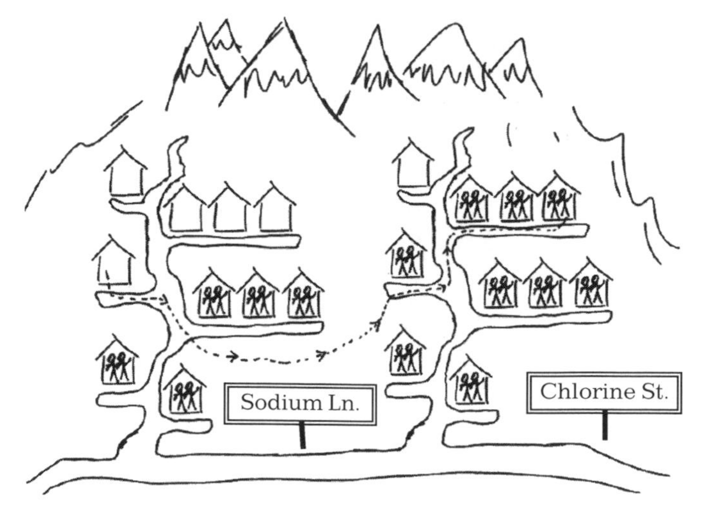
Sodium Ln.
Chlorine St.

Chlorine and sodium are both pretty nasty substances. Yet when chlorine and sodium come together the unpaired electron in sodium joins the unpaired electron in chlorine. Quantum mechanics tells us that the new electron configuration requires less energy. Now that the electron has been exchanged, the sodium atom has a net positive charge, and the chlorine atom has a net negative charge. Because of the unlike charges, the two atoms are attracted to one another and bonded together. The end result is what we know as table salt!

In table salt, an electron is completely exchanged between the atoms. Another type of bond between atoms comes about when the two atoms share electrons, creating a negative region between the two positive nuclei that attracts the nuclei toward each other.

> The underlying physical laws necessary for the mathematical theory of a large part of physics and the whole of chemistry are thus completely known.
>
> — P.A.M. Dirac, on the development of quantum mechanics

Chapter 10
Quantum Weirdness

It's amazing that quantum mechanics can tell us so much about how chemistry works. *I* might need to learn some quantum mechanics. That way I'll know if I have good chemistry with a girl before I ask her out!

Sorry Surfer Dude. We aren't talking about that type of chemistry. You'll just have to keep on taking your chances.

Quantum mechanics is the key to our scientific understanding of the structure of the atom, basic chemistry, and the interaction of light with matter. But quantum mechanics is more than a pathway to understanding the atom—it is a tool scientists have used to discover and understand things about the universe that are far more bizarre than the secrets of the atom.

Oh good! I love bizarre things almost as much as I love pudding!

The basic concepts used to determine how objects move under the influence of a force were set down by Sir Isaac Newton in 1687 in a book called *Philosophiae Naturalis Principia Mathematica*.

Newton put forth three laws of motion that form the basis for the calculations that are used by physicists and engineers to determine how to build buildings and bridges, launch rockets, reconstruct car accidents, lob mortars, and on and on—anything that involves objects and forces. The ideas formulated by Newton constitute the core of what physicists call *classical mechanics*. Newton's theory of classical mechanics is enormously successful. Buildings stand and planes fly and missiles hit targets because the theory works so well. In fact, Newton's classical mechanics can be used to understand and describe all that you encounter in everyday life.

The universe is deterministic!
Whatever that means ... heh, heh.
RADAR

According to classical mechanics, the universe is *deterministic,* which means that if you know the location and velocity of an object as well as the forces acting upon it, you can calculate exactly where it will be in the future.

One aspect of determinism is that—according to classical mechanics—it is possible to determine the *exact* location and velocity of a particle. For example, you could bounce light off a spot on a car to measure the exact location of the car, and you could bounce light off the car at two different times to figure out how the car has moved with time and from that determine the velocity of the car. Supposedly, according to Newton, as you improve your instruments and technique your determinations of position and velocity become better and better . . . with no limit on how good they can be.

Carrying the idea of a deterministic universe to an extreme, if we know the location and velocity of each particle in the universe, as well as the forces upon each particle, we could, in principle, calculate the future. It would take one mongo computer to do it. But, practicalities aside, it could be done.

That picture of the Newton guy looks familiar. Love that hair. Did he play drums for a rock band? Say maybe the Rolling Stones? Died in 1726. Didn't the Stones cut their first single before then?

Ah! The relativistic time warp of youth. Now *there's* a subject for a paper.

The universe is NOT deterministic!

$$i\hbar\frac{\partial}{\partial t}\Psi(\vec{r},t) = -\frac{\hbar^2}{2m}\nabla^2\Psi(\vec{r},t) + V(\vec{r})\Psi(\vec{r},t)$$

Dude! That's some cool graffiti. Are you in a gang?

Waves and particles are different.

If you go to a baseball game, it's easy to pick out the baseball. You can talk sensibly about the location of the baseball. Your friends might think you are weird, but you can do it. You can even specify the position of the center of the ball or the leading edge of the ball very precisely if you wish.

Suppose you go to the beach and a friend asks you to locate a wave. What would you say? It's easy enough to see the wave and point to it. The crest of a typical wave at a beach might be something like 30 meters long and 5 meters wide. But suppose you try to specify the location of the wave more precisely. Where is the center of the wave? Where is the edge? Locating a wave is not as straightforward as locating a baseball.

Classical physics deals with the particle aspect of particles while quantum mechanics treats particles as waves. The two approaches give very different results for tiny particles, where the wave aspect is important.

In other words, quantum mechanics may seem a little odd since human intuition about particles is more . . . well . . . particle-like.

Just how different are waves and particles? Imagine viewing a big seawall from above. Big ocean waves are hitting the seawall on one side. There is a small opening in the seawall through which a boat can pass. A boat passing through the opening moves in a straight line as long as the captain keeps the rudder straight. The boat acts like a particle and classical physics can predict where the boat will hit the shore.

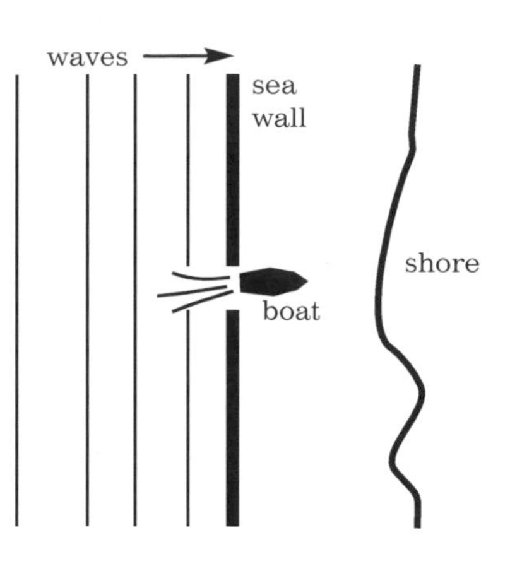
waves
sea wall
boat
shore

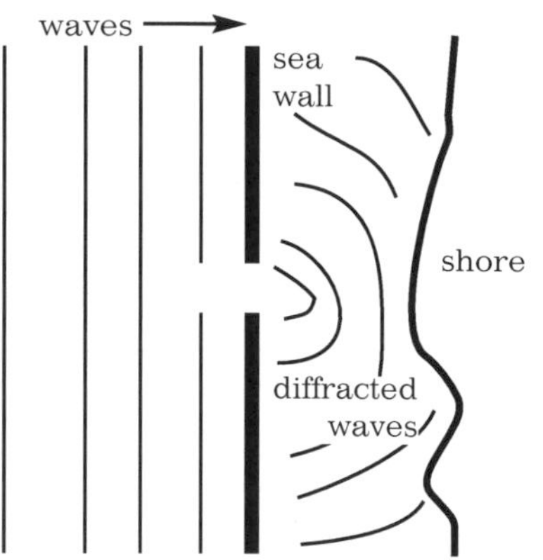
waves
sea wall
shore
diffracted waves

Waves hitting the opening in the wall in can also pass through to the water beyond. But they spread out as they go through the opening. This phenomenon is called *diffraction,* and it happens with all waves. The spread-out wavefront hits all along the shoreline.

Since particles, such as electrons, have a wavelike character, they should exhibit diffraction or spread out when they pass through a very narrow slit. Experimentally, scientists have shown that they do exactly that! This is one of the ways that we know de Broglie was correct when he postulated that particles have wavelike properties.

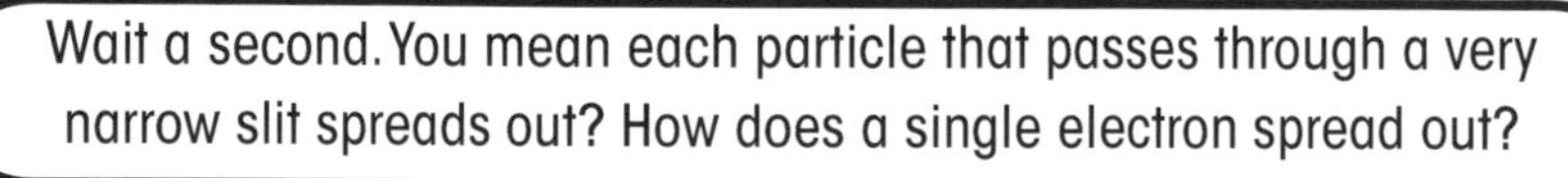
Wait a second. You mean each particle that passes through a very narrow slit spreads out? How does a single electron spread out?

Ah! Now we are getting to the core of the matter. Pardon the pun. Heh, heh. Anyway, the answer to your question is no. Quantum mechanics does not say the electron spreads out. What comes out of the quantum mechanical calculation is a solution for the electron's wave-function. What spreads out is the *wave-function.* The wave-function gives us a *probability distribution* for where the electron will be found.

Actually, I think it is the *square* of the wave-function that provides the probability distribution of where the electron is likely to be found. But maybe that detail isn't so important here.

Well, Herr Professor, if confusion is the first step to knowledge, I must be a genius. How about giving us a better feeling for this wave-function crap.

Imagine an electron passing through a thin slit and then hitting a sheet of sensitive film. The electron is a particle, and it will cause a small dark spot to appear on the film where it hits. Quantum mechanics cannot tell us where any individual electron will hit. Instead, quantum mechanics provides information about the electron wave in a form we call the wave-function which, in turn, specifies the relative probability of the electron hitting different places on the film. If we shoot 1,000 different electrons at the film, they will be distributed on the film in exactly the fashion predicted by the wave-function as determined by the quantum mechanical calculation.

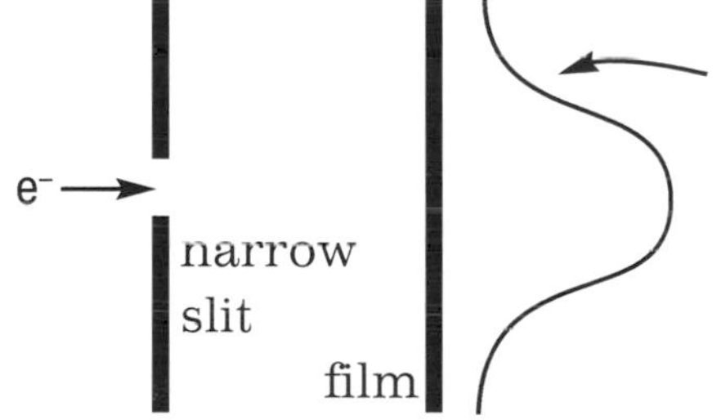

e⁻
narrow
slit
film

Probability distribution for where a single electron will hit the film. Determined from quantum mechanics.

After 1 electron passes through the slit

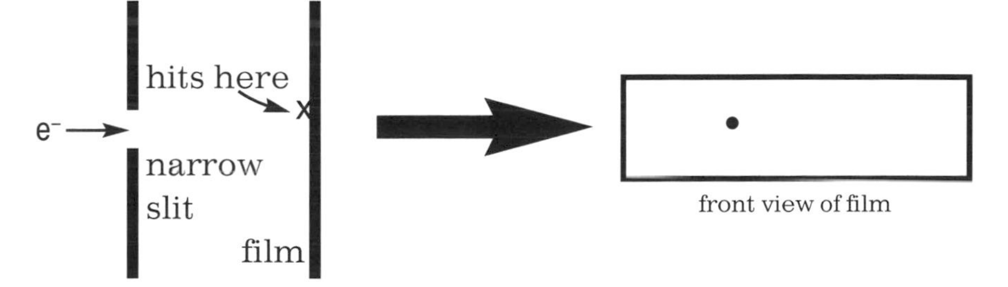

hits here
x
e⁻
narrow
slit
film
front view of film

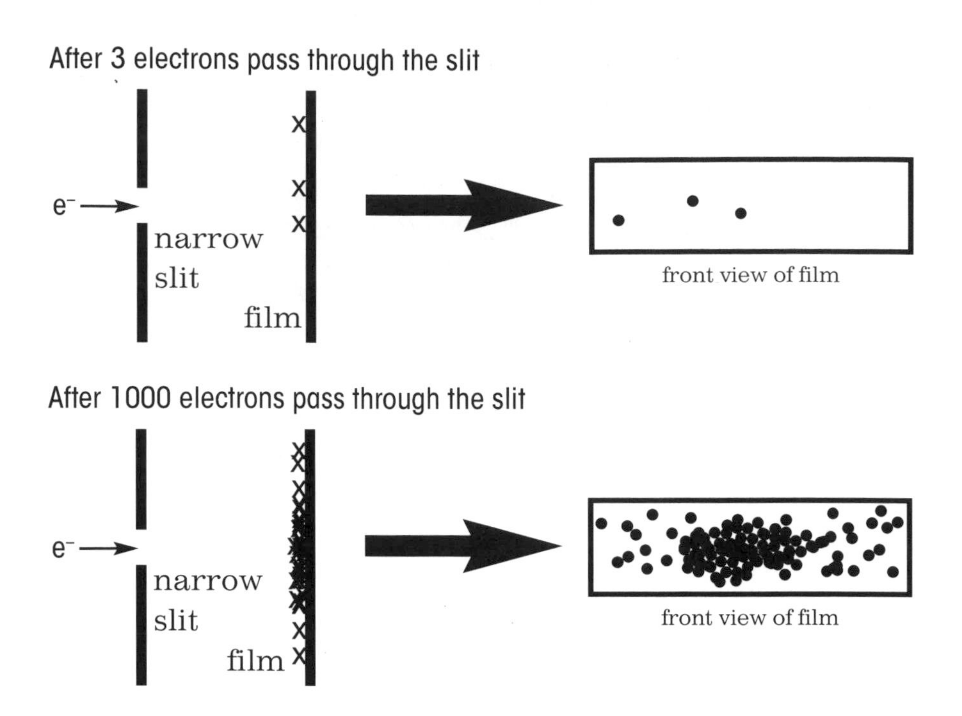

After 3 electrons pass through the slit
e⁻
narrow slit
film
front view of film
After 1000 electrons pass through the slit
e⁻
narrow slit
film
front view of film

The distribution of many electrons on the film agrees very well with the quantum mechanical calculation of the probability distribution. Still, quantum mechanics cannot tell you where any individual electron will go. Quantum mechanics only gives the relative probability of the places it might go. Once the electron hits the film, however, you know with 100 percent certainty where the electron was. This transition from a state of uncertain, probabilistic knowledge to the actual certain final state is often called the *collapse of the wave-function*.

Even if we know the location and motion of a tiny particle very well and we also know the forces acting on it, quantum mechanics says at best we can calculate the probability that the particle will take a certain path. Within all the possible paths we cannot know with certainty what will happen. This means *the deterministic universe is dead!* We cannot say for sure what will happen with individual particles.

Even if we know the location and motion of a tiny particle very well and we also know the forces acting on it, quantum mechanics says at best we can calculate the probability that the particle will take a certain path. Within all the possible paths we cannot know with certainty what will happen. This means the deterministic universe is dead! We cannot say for sure what will happen with individual particles.

This collapsing wave-function thing really bothers me because ***the universe is local!***

Al, man. I know about local buses and local restaurants, but I never heard of *local universes*. What are you going on about?

Many people through the years, including our friend Einstein, have been greatly disturbed by the concept of the collapsing wave-function in quantum mechanics. As Albert says, we live in a local universe. That's physics-speak for "we live in a world of cause and effect." Information can travel only as fast as the speed of light. That means if some event happens it can only cause another particular event to happen if there is time for a light signal to travel from the first event to the place of the second event. Otherwise, there is no way for the second event to be 'aware' that the first event happened.

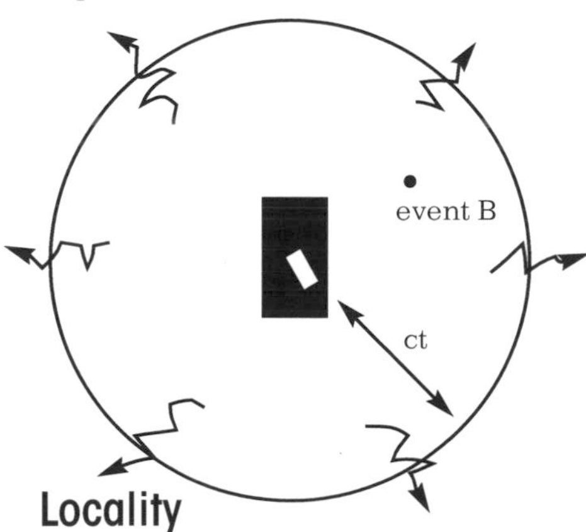
light emitted by event A
event B
ct
Locality

A switch is thrown t seconds in the past. In that time light has travelled a distance ct from the location of switch.

event C

Event B could be caused by the thrown switch theoretically.

Event C could not be caused by the thrown switch since there has not been enough time for a light signal to travel from the switch to location of event C.

The universe is NOT local!

A common interpretation of quantum mechanics (often called the *Copenhagen interpretation*) says that an electron passing through the slit destined to hit a sheet of film could show up anywhere on the film where the wave-function exists. It is only the instant that the electron is observed by causing a dark spot on the film in a specific location that the wave-function collapses, because at that instant the location of the electron is known with 100 percent certainty. In this picture of reality, the collapse of the wave-function is instantaneous. Since the wave-function has a finite size, this means the information about the collapse of the wave-function must travel faster than the speed of light. In other words, the collapsing wave-function idea seems to violate our idea of cause and effect. Proponents of this view of quantum mechanics believe the universe is *not local*—at least not at the quantum level.

This chapter is entitled "quantum weirdness" you know. There's really nothing so strange as reality, it seems. Erwin Schrödinger devised an interesting example to explore the strangeness of quantum mechanics. It is usually called *Schrödinger's* cat.

So, like where did all these quantum physics geeks go to school? Hogwarts?

Time out! You can't talk about Schrödinger's cat until you say something about radioactive decay.

Sure. I'm getting to that. Can't you go find some ballgame to ref?

Some atomic nuclei are naturally unstable. They will break down into other, more stable nuclei by emitting an electron, photon, or alpha particle.

These are known as naturally radioactive substances. Since the decay of each nucleus is a quantum mechanical process, it is impossible to know when any individual nucleus will decay.

Although it is impossible to predict when a single radioactive nucleus will decay, scientists have figured out how to quantify the decays in bulk.

Each type of radioactive substance decays with a characteristic time called a half-life. In a time of one half-life, half the nuclei in a big sample will decay. In the next half-life, half of the remaining nuclei will decay, leaving a quarter of the original nuclei. This process continues until there are no nuclei left to decay. Half-lives can be very short or very long, depending on the type of atom (nucleus).

Now, with that out of the way, I'd like you to meet Schrödinger's cat.

Initial radioactive sample
Same radioactive sample after a time of one half-life: half the nuclei have decayed.
Oh! What a sweet kitty.
Yes, he is. Now let's shove him into a box with a vial of deadly poison and a radioactive nucleus.
Don't try this at home!

Suppose we put our sweet kitty in a closed, opaque box. Also in the box is one radioactive nucleus with a half-life of one hour and a vial of deadly poison. In addition there is a radiation detector attached to a hammer. If the radioactive nucleus decays, the detector will detect the radiation and drop the hammer on the vial of poison, killing the cat.

Excuse me. I'm with the ASPCA. I heard someone is being cruel to a cat around here. But all I see is a box.

No worries. This is a thought experiment—scientific pretend. The cat is just a drawing.

The question is this: One hour later, is the cat dead or alive in the closed box? According to our understanding of radioactive nuclei, after one hour there is a 50 percent probability that the nucleus has decayed. If the nucleus has decayed the cat is dead. If not, the cat is alive.

According to a common interpretation of quantum mechanics (yes, the Copenhagen interpretation again) you can't know the exact state of the radioactive nucleus without looking inside the box. All you can know is the probability of whether it has decayed (50:50). The instant that you look in the box the wave-function collapses and the nucleus has either decayed or not, with no uncertainty. Before you open the box, a quantum physicist would say the nucleus is in a state that is a superposition of decayed and not decayed that looks is something like this:

nucleus quantum state = ½(decayed) + ½(not decayed)

Well, nuclei are nuclei and it's hard to get too upset about a confused nucleus. But the implication of the nuclear state being a superposition of decayed and not decayed is that the cat is a superposition of alive and dead like this:

$$\textbf{cat state} = \tfrac{1}{2}(\textbf{dead}) + \tfrac{1}{2}(\textbf{alive})$$

Come on! That's crazy. The cat's either dead or alive. It CAN'T be both.

Hey man! What's the big deal. After a big night on the town I'm always half dead and half alive the next day! I just never knew it was due to quantum mechanics.

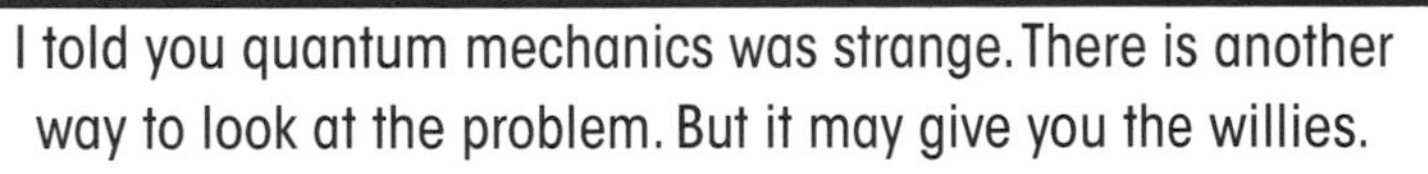

I told you quantum mechanics was strange. There is another way to look at the problem. But it may give you the willies.

Many physicists have been unsettled by the idea of an instantaneously collapsing wave-function and bizarre superposition states like a half-dead cat. In an attempt to work around these problems, Hugh Everett proposed what has come to be called the *many-worlds* interpretation of quantum mechanics in 1957.

In the many-worlds picture of quantum mechanics, the moment the cat is closed in the box the universe splits into two universes: one in which the cat is alive, and one in which the cat is dead. Now there is no need to collapse the wave-function or have a half-dead cat. But, it comes at the cost of the idea of multiple, noncommunicating, parallel universes.

Anyone who is not shocked by quantum theory has not understood it.
—Niels Bohr

Quantum mechanics is certainly imposing. But an inner voice tells me that it is not yet the real thing. The theory says a lot, but does not really bring us any closer to the secret of the Old One. I, at any rate, am convinced that He does not throw dice.
—Albert Einstein, in a letter to Max Born in 1926
(often paraphrased as "God does not play dice with the universe")

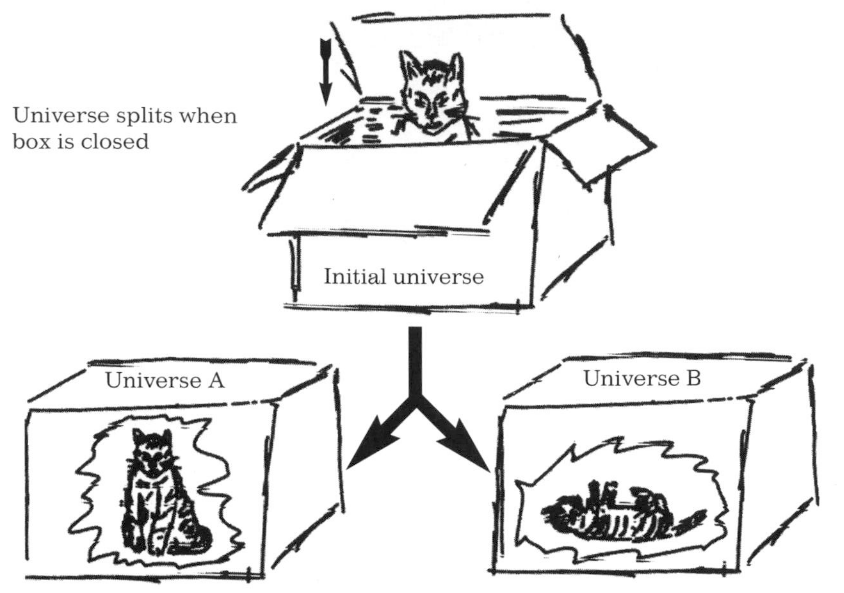
Universe splits when
box is closed
Initial universe
Universe A
Universe B

The many-worlds idea requires that the universe split so that a universe for each quantum possibility for every quantum process that happens. There are revisions of the theory that can reduce the number of universes, but still, it gets out of hand fairly quickly!

Quantum mechanics works extremely well in terms of explaining what we see in the world around us. It is not necessary that the theory agree with human intuition and the human mental comfort zone to be useful.

You mean there could be an infinite number of Surfer Dudes running amok in parallel universes?! Groan! One is bad enough.

Scientists are still working actively on trying to understand these strange issues with quantum mechanics. Stay tuned.

And if half-dead cats aren't weird enough for you . . .

The universe is really, really NOT deterministic!

$$\Delta x \Delta p \geq \frac{h}{4\pi}$$

Well, get a load of that! It's the Heisenberg uncertainty principle! You don't see *that* every day.

In the mid-1920s, Werner Heisenberg discovered there are fundamental limitations to what we can know about the universe. He found that, in the quantum world, certain pairs of quantities cannot be determined arbitrarily well simultaneously.

Well now. THAT's crystal clear. Wanna run that by me again?

Recall that in classical physics—the physics of our intuition—it is possible to determine the position and velocity (or momentum) of an object simultaneously as well as you wish.

Heisenberg found that viewing particles as waves has a very interesting consequence. It is known for waves of all types that if the crest of a wave is very narrow—which means the wave's position is well known—then the momentum of the wave could be any of a wide range of values, any of which might turn out to be what you would measure.

RADAR

Conversely, if the momentum of a wave is fixed, or known to be within a small range, then the wave crest will be very broad, and the wave's position will be poorly known. It doesn't matter how good is your laboratory equipment or technique, once a wave crest is broad enough you can't measure its location well.

Heisenberg formulated this uncertainty principle for particles. What it says is that our uncertainty in the position of a particle multiplied by our uncertainty in the momentum of a particle is greater than a tiny, but finite number given by Planck's constant divided by 4π. If we know very well where a particle is located, the momentum must be very uncertain to keep the product of the two uncertainties larger than the specified value. Conversely, if the momentum of the particle is known very well, the location of the particle will not be well determined.

In other words, if THIS one is big, the other is small.
Δp
Δx
Δx
And, if THIS one is big, the other is small.
Δp

Imagine an electron zipping past you with a specific speed (think of this as momentum if you want). Your job is to measure both the location and the speed of the electron as precisely as possible simultaneously.

That'd be easy enough to do, right? Just as we might use a laser beam to determine the position and velocity of a car, we could bounce a photon or two off the electron to get at the location and speed of the electron.

Yeah. And because the wavelength of the photon gets smaller if it is more and more energetic, we can locate the electron arbitrarily well by using more and more energetic photons.

The problem with this idea is that hitting a tiny electron with an energetic photon is sort of like hitting a car with another car. It blasts the electron and leaves it with an uncertain momentum. If you hit the electron with a more energetic photon in order to determine its location ever more precisely, it blasts the electron even more and the uncertainty in the electron's momentum is even greater!

$$\Delta x \Delta p \geq \frac{h}{4\pi}$$

Light can bounce off a car and not affect the car's speed or momentum in any noticeable way. Energetic light scattering off an electron, however, will affect the electron's motion, leaving the momentum uncertain. This is an example of Heisenberg's uncertainty principle in action.

It seems that Heisenberg has done away with the deterministic universe for sure. Even if you had the ultimate mongo computer, you could not determine the location and motion of every particle perfectly in order to input it into the computer to calculate the future. For better or worse, we are destined to live in an unpredictable world at the quantum level.

> I think it is safe to say that no one understands quantum mechanics.
>
> —Richard Feynman

Werner Heisenberg goes out for a drive and is pulled over by a traffic cop. The policeman gets out of his car, strolls up to Heisenberg, and asks, "Sir, do you know how fast you were going?"

Heisenberg looks at the cop blankly and says, "No. But I know *exactly* where I am."

Chapter 11
Quantum Weirdness Meets the Universe

The very small and the very large are intimately related in our universe!

$$\Delta E \Delta t \geq \frac{h}{4\pi}$$

There's more?! Dude, my head blew up about 15 pages ago. And to think, I've always been a fan of waves.

Well, dang! It's another version of Heisenberg's uncertainty principle. Wait'll the guys back at the station hear about this!

Recall that Werner Heisenberg discovered that, in the quantum world, certain pairs of quantities cannot be determined arbitrarily well simultaneously. One pair of such quantities, position and momentum, leads to the downfall of quantum determinism. There are other pairs of quantities that can be formed into an uncertainty principle. For example, a different form of the quantum uncertainty principle involves energy and time. Hold on to your hat, because much weirdness springs from this particular form of uncertainty principle.

You can think about the energy-time uncertainty principle like this: If a particle exists in a certain quantum state for a very short time, its energy is very uncertain.

That doesn't sound weird. It's sad. The poor particle is confused because it's temporally deprived. We need to love and nurture all particles, don't you think?

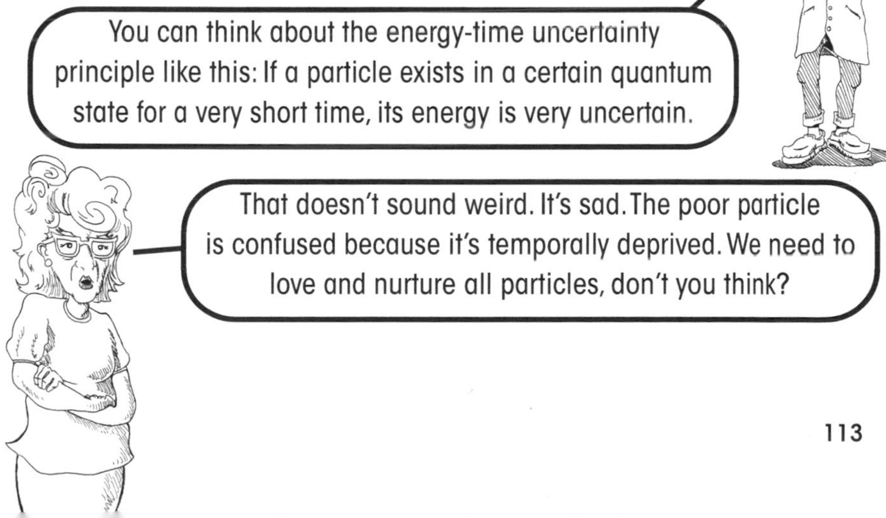

And now presenting, for your viewing pleasure . . . er . . . nothing!

To really get an idea of how bizarre the quantum world can be, let us consider absolutely nothing. Yep, nothing—or what a physicist might call *the vacuum*. In classical physics the vacuum consists of nothing at all—unchanging, boring nothing. It's difficult to imagine a box of absolutely nothing—not even air—yet that's what a scientist means when she talks about a vacuum.

In the world of quantum mechanics, however, things are a little different. Recall that quantum mechanics isn't so much a way of looking at exactly what happens in a specific case, but rather all the possibilities and probabilities of what could happen. Things that *can* happen *will* happen.

Oh, come off it. It's not so hard to imagine "nothing." Just step into Surfer Dude's head!

Okay. But, what does that have to do with nuthin'?

Imagine an empty bit of space—a vacuum. Quantum mechanics allows the energy in a region of space to vary, or fluctuate, in a way that is consistent with the uncertainty principle. In other words, even in a box of nothing the energy can fluctuate to very large values so long as it happens over a very short time. If Δt is tiny, ΔE can be big. These energy variations allowed by the uncertainty principle are called quantum fluctuations. Remember mass-energy equivalence? If the energy fluctuation is large enough the energy can turn into a particle-antiparticle pair. The pair disappears almost instantly, because it can only hang around momentarily if the uncertainty principle is to remain inviolate.

The upshot is that the vacuum is anything BUT empty. It is a seething sea of particle-antiparticle pairs popping in and out of nothing. The particle pairs hang around for such a short time that you could never see them. Still, sophisticated scientific experiments have uncovered solid evidence of this *quantum vacuum*.
e^+e^- qq $q\bar{q}$ e^+e^-
e^+e^-
$q\bar{q}$ Box of nothing ... $q\bar{q}$
e^+e^-
Wow! And people call *ME* weird.
Hold on you furry-faced, pencil-necked geek! You thought you could sneak one by me, heh? What do you mean when you say "antiparticle"? That's like Star Trek gibberish.
Ah. Yes. Sorry. Every type of fundamental particle has an antiparticle that is exactly like its particle partner but has the opposite electric charge. You've heard of the electron, with its negative electric charge? Well, scientists have discovered (and regularly make) positive electrons, positrons, in the lab. Scientists have even made anti-protons and anti-hydrogen atoms.
So there's an anti-me someplace?
There's a scary thought.

If there was enough antimatter around, the particles could join together into antimatter atoms and perhaps antimatter dogs and so forth. The thing is, whenever antimatter meets matter, the particles annihilate each other and the mass all turns into energy in the form of high-energy photons.

Why is it the universe is all matter then? Why is it not antimatter or a mix of matter and antimatter? If I ate anti-lasagna, would I lose weight?

Nobody knows why the universe is matter rather than antimatter. Scientists have discovered small differences between matter and antimatter and are investigating those differences to see if they explain the predominance of matter in the universe. That's still a puzzle on which scientists are working.

You know, I'm bored with all this talk about nothing and *anti-things*.

Me too. Let's talk about the nature of forces in quantum mechanics instead!

The uncertainty principle plays a critical role in our modern understanding of the particles and forces of nature. Here's how it works: imagine an electron moving through space . . .

e^-

Suppose this electron emits a photon that turns into an electron-positron pair that then turn back into a photon that is reabsorbed by the electron.

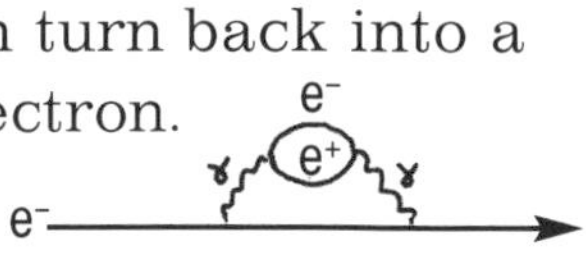

e^- e^+ γ γ e^-

You CAN'T do that! It breaks energy conservation. You can't get something from nothing!

Sure it can happen . . . so long as it happens over such an extremely short time that it doesn't violate the uncertainty principle. Welcome to the wacky, magical world of quantum mechanics!

Quantum mechanics views a particle as a sum of all the possibilities allowed to happen by the uncertainty principle, charge conservation, and so forth. So in the quantum world an electron is not like a tiny marble but rather more like a fuzzy cloud of "virtual" particles that pop in and out of existence in a time too short to observe.

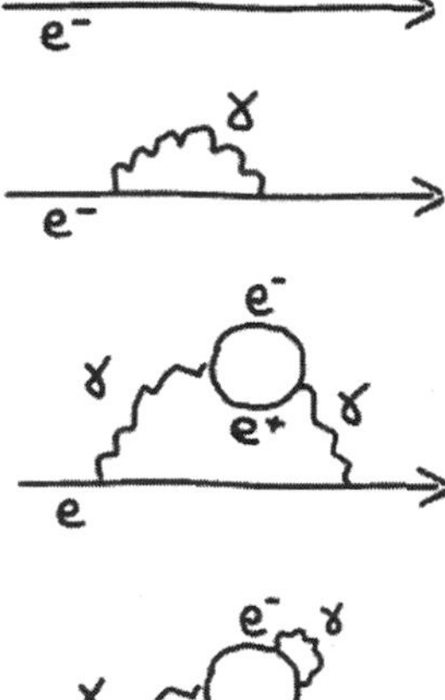
e⁻
γ
e⁻
e⁻
γ
e+
γ
e

e⁻
γ
γ
γ
e+
γ
e
γ

I see the electron as the sum of all these different possibilities plus others.

Hey sport. Those are pretty pictures alright. Looks like some stuff I saw in a tattoo shop in LA. But, seriously now, why should I give a crap about particles that can't exist long enough for me to see 'em? I mean all these so-called virtual particles come and go, so who cares?

Imagine you have amazing superpowers and can stop time and see subatomic particles. Suppose you see an electron moving through space. At any given moment, if you froze time and inspected the electron closely, you would find that it is made up of a cloud of virtual particles which normally come and go so fast you can't observe them. Quantum mechanics can't tell you which virtual particles will be present, only the probability that you'll find particular configurations of virtual particles in the cloud. The virtual particles in the cloud carry the essence of the original electron. In other words, in total, the particles in the cloud carry the electric charge and the momentum of the original particle.

Cute tights! Did you notice the hair on your legs sticking through the material? That's sort of gross.

PARTICLE MAN

Electron cloud of virtual particles frozen in time by Particle Man

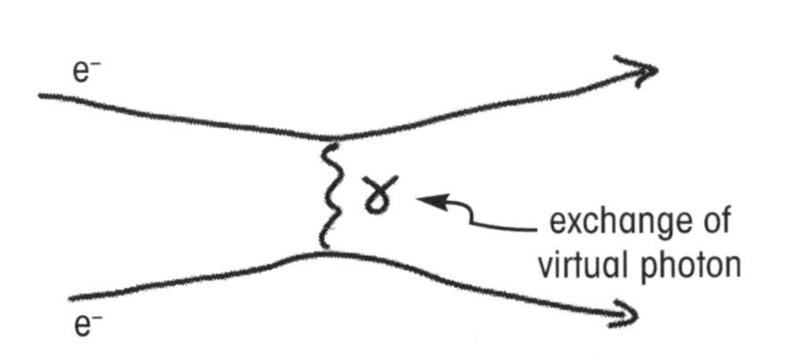
e⁻
γ
exchange of
virtual photon
e⁻

Now, rather than stopping time, you let it move forward slowly while the electron you are watching passes near another electron. You observe something very odd. One of the virtual particles gets exchanged between the two electron clouds. Since that virtual particle carries momentum with it, you have witnessed the particles exchange momentum. That is *the essence of a force*.

Imagine two skaters on ice tossing a heavy workout ball back and forth. When one skater tosses the ball, she moves backward. When the other skater catches the ball, he moves backward. The exchange of the ball is like a force between the two skaters.

In quantum mechanics, the essence of a force is the exchange of virtual particles. The nature of the force is determined by the type of virtual particle that is exchanged.

Hey. My last date called me a one-track-minded force of nature. Guess that means I've got some cool virtual particles popping up here and there.

As wacky as this idea of a force may seem, it works extremely well, and it underlies the fundamental theory of forces and the structure of matter known as . . .

the Standard Model!

The forces you observe in everyday life can be traced to the fundamental forces of gravitation or electromagnetism.

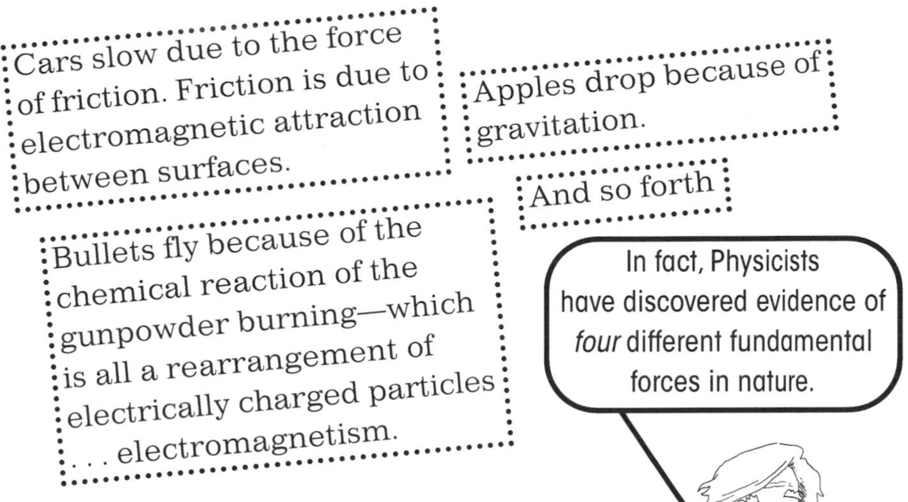
Cars slow due to the force of friction. Friction is due to electromagnetic attraction between surfaces.
Apples drop because of gravitation.
And so forth
Bullets fly because of the chemical reaction of the gunpowder burning—which is all a rearrangement of electrically charged particles . . . electromagnetism.
In fact, Physicists have discovered evidence of *four* different fundamental forces in nature.

Strong nuclear force

The strongest force known in nature holds together the nucleus and binds together the particles, known as quarks, that make up protons and neutrons. This force has a range that is no larger than an atomic nucleus.

Electromagnetism

This force is 20 to 50 times weaker than the strong force. Electromagnetism is responsible for holding electrons in atoms and molecules. All chemical reactions are electromagnetic in nature. All interactions with light at any wavelength are due to the electromagnetic force.

Weak nuclear force

This force is almost the same strength as the force of electromagnetism, but weak nuclear interactions rarely happen. This force is responsible for some types of radioactive decay

Gravitation

This force is, far and away, the weakest force known in nature. In general relativity this force is thought of as being due to a warping of space-time. Many physicists believe that someday we will also understand gravity through a theory that views the force as an exchange of virtual particles, but we do not have a working theory of quantum gravitation yet.

Sorry Surfer Dude. Looks like you didn't make the list as a force of nature.

The fundamental particles

In the Standard Model, there are three different types of fundamental particles in the universe: quarks, leptons, and gauge bosons.

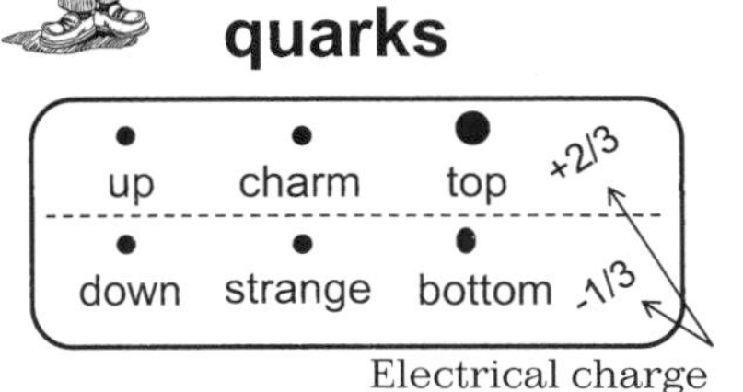
quarks
up
charm
top
+2/3
down
strange
bottom
-1/3
Electrical charge

There are six types of quarks. These particles possess fractional electric charge and combine in twosomes or threesomes to create other particles like protons and neutrons. The strong force is the glue that holds quarks together.

There are six types of leptons. The most familiar lepton is the electron. The muon and tau are just like the electron only much heavier. The electron, muon, and tau possess electric charge and can interact with other particles via the electromagnetic interaction. There are also three leptons known as *neutrinos* that are very nearly massless and carry no electric charge. They feel only the weak nuclear force. As such, they rarely interact with other particles.

leptons

electron	muon	tau	-1
electron neutrino	muon neutrino	tau neutrino	0

gauge bosons

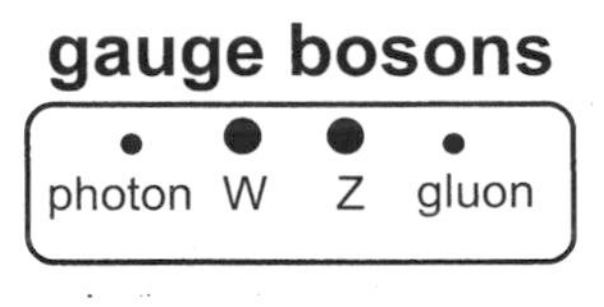
photon
W
Z
gluon

The gauge bosons are the particles that get exchanged to convey the force. The electromagnetic force is conveyed when a virtual photon is exchanged, the strong force is carried by the gluon, while the weak nuclear force is conveyed by the W and Z particles.

Higgs

Another particle, the *Higgs particle,* is an important part of the Standard Model. The Higgs has not yet been seen in an experiment, but the search continues.

You're not serious, right? I mean what were you physics twits thinking when you made up these names?
Heh, heh, heh. I'm just trying to imagine how a scientist studying "strange" or "bottom" particles explains his life work to his mom.

Most of the particles in the Standard Model have been seen by physicists using powerful particle accelerators that smash beams of particles together. The energy available after the collision turns into particles that decay and/or stream away from the collision, where they are observed by large detectors. With this technique, scientists have been able to study quantum particles and processes that are never seen normally, but that are essential for a full understanding of how nature works and how the universe has evolved into what it is now.

Oh PUH-LEASE! How can stuff too small to see affect something unimaginably large like the universe itself? You must have quarks on the brain or something like that.

The push into inner space has been very exciting and led to many Nobel Prizes in Physics. A good place to learn more is http://www.particleadventure.org/.

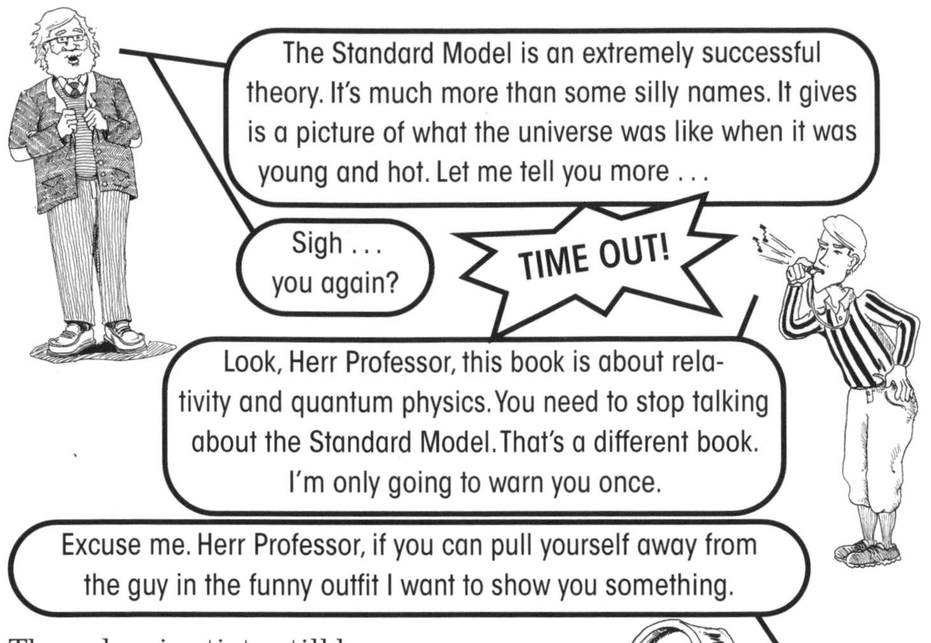

The Standard Model is an extremely successful theory. It's much more than some silly names. It gives is a picture of what the universe was like when it was young and hot. Let me tell you more . . .
Sigh . . . you again?
TIME OUT!
Look, Herr Professor, this book is about relativity and quantum physics. You need to stop talking about the Standard Model. That's a different book. I'm only going to warn you once.
Excuse me. Herr Professor, if you can pull yourself away from the guy in the funny outfit I want to show you something.

Though scientists still have many, many interesting and puzzling questions to explore, they understand a great deal about the structure of matter and the forces of nature through quantum mechanics and the Standard Model.

One of the most exciting things to come out of this new understanding is the realization that the quantum world of the very tiny is intimately related to the large-scale structure and history of the universe.

It seems that all the faraway galaxies I see are moving away from us. Isn't that strange?

I was about to say something about that before that strange referee interrupted me. You see, we live in an expanding universe.

Edwin Hubble (1889–1953) The Hubble space telescope was named in his honor. No doubt he had happier moments in his life than he appears to be having in this photograph.

In the first part of the 1900s, astronomers noted that, except for the closest galaxies, all the galaxies in the universe appear to be moving away from us. In 1929, Edwin Hubble and Milton Humason observed that the speed with which the galaxies are receding from us increases with the distance of the galaxy from the Milky Way. This is strong experimental evidence that the space in our universe is expanding!

The galaxies are moving away from us. Okay. But I don't understand why scientists think that means the space in the universe is expanding. Why is that?

It's sorta like our galaxy has bad cosmic BO, right?

Imagine watching raisin bread dough rise. In the initial dough, the raisins are mixed uniformly. As the bread rises, the dough between the raisins expands. Each raisin moves apart from the others. If you consider the view from one of the raisins, however, you see that nearby raisins move away less as the bread rises than the raisins that started out farther away. The expansion of space in the universe, in analogy to the expanding dough, leads to precisely the result noted by Hubble and Humason.

Oh, this is so exciting! I just LOVE astrology and cosmetology!

Er . . . uh . . . Do you mean *astronomy and cosmology?* That's what we're talking about. Cosmetology is hair-dos and pretty nails. Cosmology is the study of the universe in its totality. We are discussing scientific cosmological ideas here.

What is getting bigger with time was once smaller.

Just like my belly!

The realization that we live in an expanding universe spawned the big bang theory of the universe.

There are other ideas consistent with experiment, and physicists are working hard—experimentally and theoretically—to get a clearer picture of the history of the universe. Here is one scenario that is popular among scientists today. It goes by the name of *inflationary, hot big bang model*.

Oh, baby! Inflationary hot big bang model. Now, *there's* a sexy name!

In order to step through the inflationary, hot big bang scenario, you'll need to invoke serious superpowers.

No problem, mate. I'll be your tour guide.
PARTICLE MAN

Suppose, with the help of your super-powerful guide, you travel back in time some 13 billion years to about the time our universe began. Where are you if the universe isn't there? Good question. That's why you needed your super-powerful friend. Let's say for now that you find yourself in another universe or perhaps some sort of space-time foam. The scientific data doesn't tell us much about what might have been present. Because quantum mechanics is part of nature, quantum fluctuations are present. They are happening all over the place. In each of these fluctuations, the nature of the virtual particles created can vary. Suddenly, a particular quantum fluctuation happens where the conditions are such that an enormous expanding pressure exists in the space-time of the fluctuation.

What caused this expanding pressure? Is that what's happening to my belly?

Even among proponents of this model in the scientific community there are different ideas about what might have caused this pressure (if it happened). There are various ways in which it might have happened, and no scientific evidence yet supports one particular choice over any other. Oh, and by the way, I suspect other things are happening with that belly.

Inflating space-time of initial fluctuation

Observable universe

Quantum fluctuations in energy blown up in size by the inflation during the expansion

Driven by the enormous expanding pressure, the little bit of space-time expands extraordinarily fast. This period of expansion is called *inflation*. The expansion is so very fast—much faster than the speed of light—that the initial quantum fluctuation becomes much larger than the *observable universe,* which is the distance light would have traveled in the same time. Since this happens in the tiniest fraction of a second after the fluctuation, the observable universe at this time is still very small by human standards.

As inflation is happening, quantum fluctuations in the local energy occur in the expanding space-time. Because the space-time is expanding so fast, these new fluctuations rapidly blow up in size and can no longer vanish, leaving the new little universe crisscrossed with regions that have slightly different energies or temperatures.

Sort of like stretch marks on the universe?

At this stage, according to the model, the universe has a scattering of little hot spots rather than markings or stretch marks. The temperature fluctuations are not large—only about 1 part in 100,000!

After a tiny fraction of a second of this hyperexpansion, or inflation, the quantum condition that caused the expanding force is changed and the energy tied up in the expansion is dumped into radiation (light) and a primordial soup of subatomic particles. The expansion of the universe continues, but at a much slower rate—nothing like the earlier period of inflation.

Subatomic particle soup

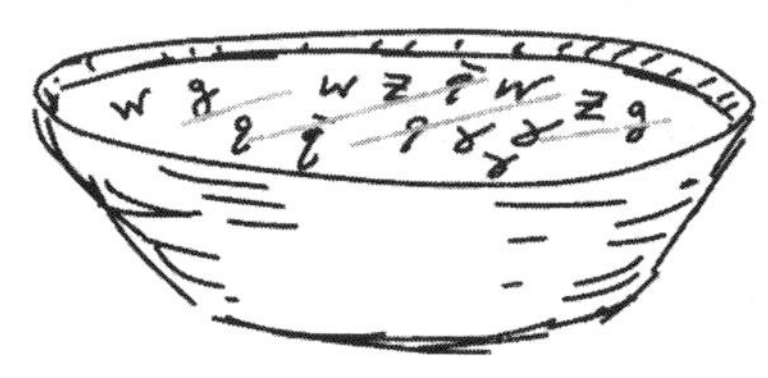

The particles and typical particle energy present in the universe in the primordial soup phase are similar to those achieved in the highest energy accelerators here on earth. At about one microsecond after the universe begins, the universe cools to a temperature about 40,000 times hotter than the center of the sun. At this temperature, the quarks in the primordial soup condense into protons and neutrons. The density of particles—the number of particles in a little volume—is higher in the regions where the universe has hot spots.

protons, neutrons, photons

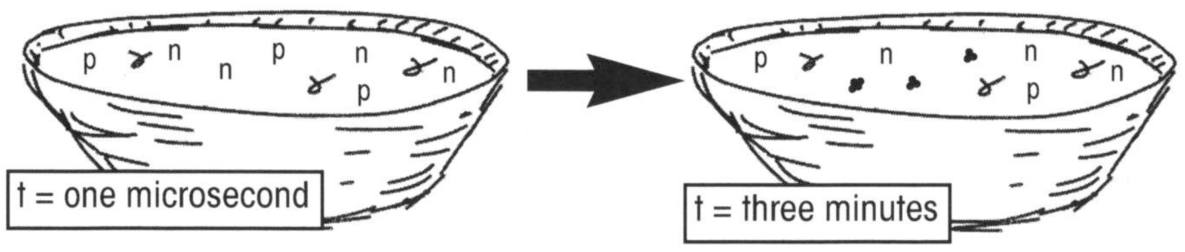
protons, neutrons, light nuclei, photons
t = one microsecond
t = three minutes

Two to three minutes after the universe begins, things cool enough so that some of the protons and neutrons stick together, forming light nuclei. The universe consists mostly of protons and electrons and photons, with some light nuclei. In this hot gas of electrically charged particles the photons can't travel very far. So, in a real sense, the universe is opaque.

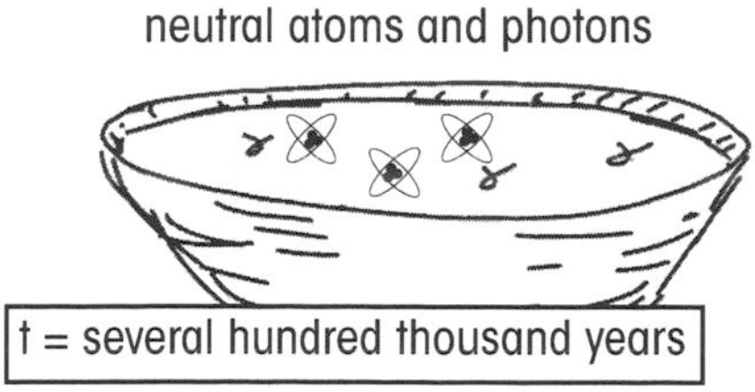

neutral atoms and photons
t = several hundred thousand years

Several hundred thousand years after the universe begins, it cools to approximately 3,000 degrees. This temperature is cool enough so that the electrons can join with the protons and light nuclei to form atoms. Now that the gas in the universe is mostly uncharged, the universe becomes transparent to light. Some of that light travels unimpeded for thirteen billion years and is seen by scientists on Earth. That light appears to come at Earth from all parts of the sky and is called the *cosmic microwave background*.

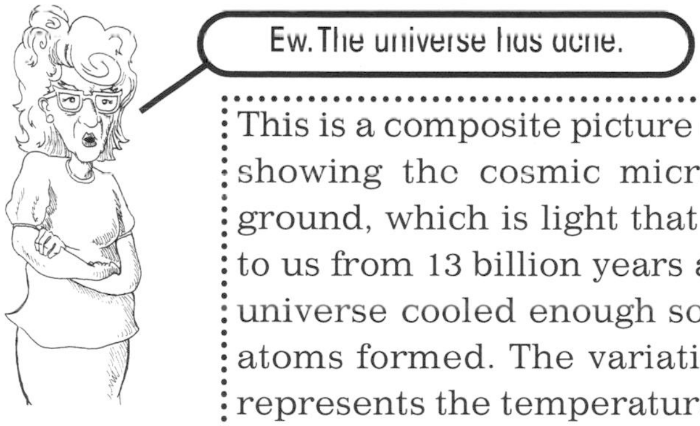

Ew. The universe has acne.

This is a composite picture of the full sky showing thc cosmic microwavc background, which is light that has traveled to us from 13 billion years ago when the universe cooled enough so that neutral atoms formed. The variation in shades represents the temperature variation in different regions of the universe at that time. This is a figure created with data from NASA's WMAP satellite.

Recall that quantum fluctuations during inflation became the hot and cold spots in the later universe. This structure provided the nonuniformity necessary to form the stars and galaxies we see today.

As time passes, the universe continues to cool. Regions of space with more atoms begin to collapse due to the force of gravity. As the gas clouds form, they become very hot. Eventually many of the gas clouds become hot enough that nuclear reactions begin in their cores and stars are born. Clumps of stars and gas become the galaxies that we see today.

> The more than 100 billion galaxies, sparkling throughout space like heavenly diamonds, are nothing but quantum mechanics writ large across the sky.
>
> —Brian Greene,
> *The Fabric of the Cosmos*

That's it buster! You're under arrest for flagrant dissemination of too much information.

What do you mean I'm under arrest? You're just a referee. How can you put me in jail?

Just think of it as a penalty box with bars.

I think . . . er, possibly . . . could it be that the meaning of life is not pudding after all? I know! The meaning of life is rising raisin bread dough and primordial soup!
All this talk of food is making me hungry. Someone holdeth my spot here. I'm going to search for meaning in a cheeseburger.
Well, whaddaya think?
I think we're about done here.
I'm tellin' you what. You two have been done for a long time. Let's finish this thing.
THE END

A few suggested websites for further exploration:

Einstein and Relativity

- http://nobelprize.org/educational_games/physics/relativity/
- http://archive.ncsa.uiuc.edu/Cyberia/NumRel/EinsteinLegacy.html
- http://www.phys.unsw.edu.au/einsteinlight/

Light

- http://science.howstuffworks.com/light.htm

Quantum Physics

- http://nobelprize.org/educational_games/physics/quantised_world/
- http://physics.about.com/od/quantumphysics/p/quantumphysics.htm

Nuclear Physics

- http://library.thinkquest.org/3471/

Particle Physics, Matter, Forces

- http://particleadventure.org/

Cosmology

- http://map.gsfc.nasa.gov/
- http://www.astro.ucla.edu/~wright/cosmolog.htm

About the Author & Illustrator

Steven Manly received an undergraduate degree from Pfeiffer College and a Ph.D. in high energy physics from Columbia University. He moved up the faculty ranks at Yale University before moving to the University of Rochester, where he now resides and terrorizes students in the introductory physics course sequences. Professor Manly works on experiments at high energy accelerators around the world where his research probes the structure of matter and the forces of nature. Recently, he was named the recipient of the 2007 Excellence in Undergraduate Teaching Award by the American Association of Physics Teachers.

Steven Fournier was born and raised in Massachusetts. Amidst the 10 years he spent carousing throughout Boston, he earned a BFA degree in Illustration and Animation from Massachusetts College of Art and Design, played music with friends, and became handy with a Chef's knife. His current projects include (but not exclusively) apprenticing to be a tattoo artist, making photocopy comic book zines, training his hand to draw in his sleep, and playing country cover songs on his guitar.

Presently, he lives somewhere between his head and Worcester, MA. He can be found on the internet at http://everydoghasitsdesign.com.

THE FOR BEGINNERS® SERIES

AFRICAN HISTORY FOR BEGINNERS:	ISBN 978-1-934389-18-8
ANARCHISM FOR BEGINNERS:	ISBN 978-1-934389-32-4
ARABS & ISRAEL FOR BEGINNERS:	ISBN 978-1-934389-16-4
ANARCHISM FOR BEGINNERS:	ISBN 978-1-934389-32-4
ART THEORY FOR BEGINNERS:	ISBN 978-1-934389-25-6
AYN RAND FOR BEGINNERS:	ISBN 978-1-934389-37-9
BARACK OBAMA FOR BEGINNERS, AN ESSENTIAL GUIDE:	ISBN 978-1-934389-44-7
BLACK HISTORY FOR BEGINNERS:	ISBN 978-1-934389-19-5
THE BLACK HOLOCAUST FOR BEGINNERS:	ISBN 978-1-934389-03-4
BLACK WOMEN FOR BEGINNERS:	ISBN 978-1-934389-20-1
CHOMSKY FOR BEGINNERS:	ISBN 978-1-934389-17-1
DADA & SURREALISM FOR BEGINNERS:	ISBN 978-1-934389-00-3
DECONSTRUCTION FOR BEGINNERS:	ISBN 978-1-934389-26-3
DEMOCRACY FOR BEGINNERS:	ISBN 978-1-934389-36-2
DERRIDA FOR BEGINNERS:	ISBN 978-1-934389-11-9
EASTERN PHILOSOPHY FOR BEGINNERS:	ISBN 978-1-934389-07-2
EXISTENTIALISM FOR BEGINNERS:	ISBN 978-1-934389-21-8
FOUCAULT FOR BEGINNERS:	ISBN 978-1-934389-12-6
GLOBAL WARMING FOR BEGINNERS:	ISBN 978-1-934389-27-0
HEIDEGGER FOR BEGINNERS:	ISBN 978-1-934389-13-3
ISLAM FOR BEGINNERS:	ISBN 978-1-934389-01-0
KIERKEGAARD FOR BEGINNERS:	ISBN 978-1-934389-14-0
LACAN FOR BEGINNERS:	ISBN 978-1-934389-39-3
LINGUISTICS FOR BEGINNERS:	ISBN 978-1-934389-28-7
MALCOLM X FOR BEGINNERS:	ISBN 978-1-934389-04-1
NIETZSCHE FOR BEGINNERS:	ISBN 978-1-934389-05-8
THE OLYMPICS FOR BEGINNERS:	ISBN 978-1-934389-33-1
PHILOSOPHY FOR BEGINNERS:	ISBN 978-1-934389-02-7
PLATO FOR BEGINNERS:	ISBN 978-1-934389-08-9
POSTMODERNISM FOR BEGINNERS:	ISBN 978-1-934389-09-6
SARTRE FOR BEGINNERS:	ISBN 978-1-934389-15-7
SHAKESPEARE FOR BEGINNERS:	ISBN 978-1-934389-29-4
STRUCTURALISM & POSTSTRUCTURALISM FOR BEGINNERS:	ISBN 978-1-934389-10-2
ZEN FOR BEGINNERS:	ISBN 978-1-934389-06-5
ZINN FOR BEGINNERS:	ISBN 978-1-934389-40-9

www.forbeginnersbooks.com